BRITISH ASCIDIANS

A NEW SERIES

Synopses of The British Fauna No. 1

BRITISH ASCIDIANS

TUNICATA : ASCIDIACEA

Keys and Notes for the Identification of the Species

R. H. MILLAR

Marine Station, Millport, Scotland

1970

Published for

THE LINNEAN SOCIETY OF LONDON

by

ACADEMIC PRESS

LONDON AND NEW YORK

ACADEMIC PRESS INC. (LONDON) LTD
Berkeley Square House
Berkeley Square,
London, W1X 6BA

U.S. Edition published by
ACADEMIC PRESS INC.
111 Fifth Avenue,
New York, New York 10003

Library of Congress Catalog Card Number: 75–85463
SBN: 12-496650-0

Printed in Great Britain by
Butler and Tanner Ltd, Frome and London

A Synopsis of the British Ascidians

R. H. MILLAR

Marine Station, Millport, Scotland

CONTENTS

Introduction

The ascidians constitute a class of the subphylum Tunicata, which are chordate animals related to the vertebrates. Their exact relationship is obscure and the chordate nature of ascidians is more apparent in the notochord and dorsal central nervous system of the larva than in any adult character. The adults of ascidians are sessile but the larvae of most species are motile forms known as tadpole larvae because of their shape.

All ascidians are marine and in places certain species are sufficiently abundant to form a conspicuous part of the fauna, both on sheltered rocky shores and in deeper water. Their identification is not always easy and in many cases requires dissection and microscopic examination of internal features.

In some species the animals are solitary, with individuals attaining a length of several centimetres, but in others colonies are formed with small zooids more or less embedded in a common matrix or joined by basal stolons at least during part of their life. The fundamental structure is similar in solitary and colonial types and may be illustrated by the common species *Ciona intestinalis* (Linnaeus).

Structure, as illustrated by *Ciona*

The body (Fig. 1) is approximately cylindrical, attached by one end or also along part of its length. At the opposite end is the *oral* or *branchial siphon* through which water is drawn into the pharynx and the *atrial siphon* from which it is expelled. The whole superficial layer of the body comprises the test, which is a protective substance chemically similar to cellulose. In *Ciona* the test is soft and thin but in some genera it is hard and thick and in the compound forms it usually constitutes a common matrix in which the *zooids* or individuals are embedded. The margin of the oral siphon has eight rounded lobes and the atrial siphon six lobes; in other genera the number of lobes may be different. Longitudinal, circular and oblique muscles are embedded in the body wall. The oral siphon leads back into the *branchial sac*, the entrance to which is guarded by a ring of finger-like *oral* or *branchial tentacles*. Although these are simple in *Ciona*, they are branched in certain families. Immediately behind the oral tentacles, on the roof of the pharynx, is the small *dorsal tubercle*, with the horse-shoe shaped opening of a narrow duct leading from the *neural gland*. This gland is applied to the lower surface of the *ganglion*. The ganglion is visible externally as a small white spindle-shaped body lying between the bases of the siphons and represents the central nervous system.

The walls of the branchial sac are perforated by transverse rows of openings called *stigmata*; these are subdivisions of the pharyngeal slits. Between adjacent rows of stigmata the walls of the branchial sac are raised to form *transverse bars*. *Longitudinal bars* are also present and at their intersections with the transverse bars they bear the small *branchial papillae*. A series of slender curved *dorsal languets* extends along the mid-dorsal line of the branchial sac. The *endostyle* is a rod-like grooved structure occupying the mid-ventral line of the branchial sac. Variations in the branchial structure of other genera include: spirally coiled stigmata, internal *folds* on the branchial walls and a continuous *dorsal lamina* in place of the series of languets.

The *oesophagus* is a curved funnel-shaped tube leading from the postero-dorsal corner of the branchial sac back to the spindle-shaped *stomach*. Diverticula are developed on the walls of the stomach in some genera and a curved *pyloric caecum* is present in many although both features are absent in *Ciona*. The *intestine* is a narrow tube curving dorsally to the long straight *rectum*, which lies dorsal to the branchial sac, and opens by the *anus*, into the *atrial cavity*. This cavity is a space between the body wall and the branchial sac and leads at its anterior end into the base of the atrial siphon.

The *heart*, situated behind the branchial sac, is a V-shaped tube. The gonads consist of the *ovary* lying between the stomach and intestine, and the branched *testis* which is spread over the surface of the intestine and part of the stomach. Some genera, however, have very different arrangements and the gonads are often situated laterally in the body wall and may number from one to many separate bodies. In *Ciona* the *genital ducts* accompany the rectum and extend beyond it to open near the base of the atrial siphon. One structure which is not found in *Ciona* but which is a conspicuous feature of some genera is the large *renal sac* on the right side of the body.

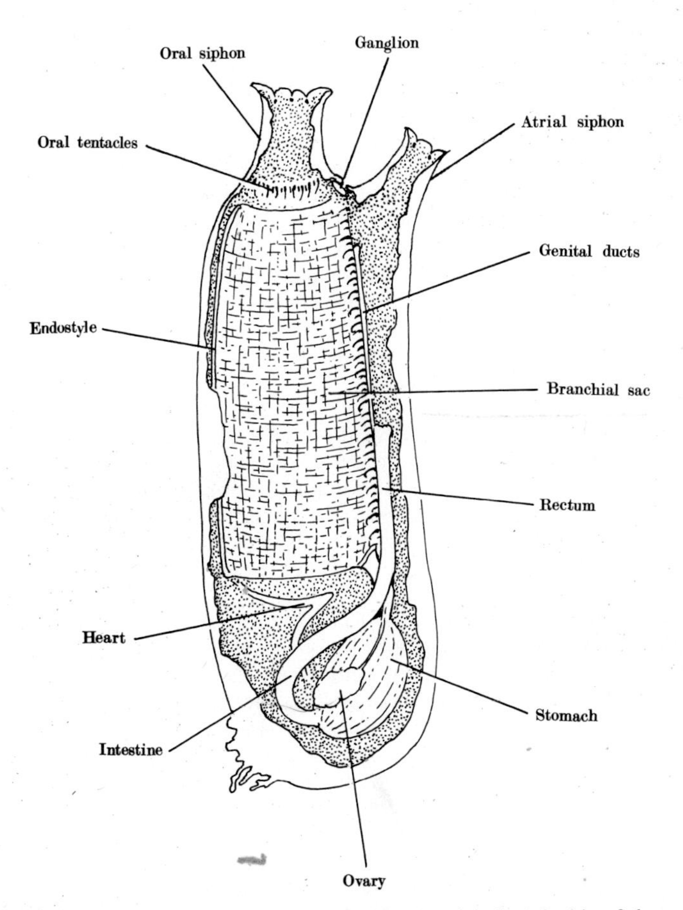

Fig. 1. *Ciona intestinalis* (L.), from the left, with the left side of the body wall and branchial wall removed.

Biology and Ecology

The group is world-wide in occurrence but most species have their own characteristic geographical distributions and within these they are restricted to particular ecological conditions. Ascidians have colonized most types of marine habitat, from rocky shores to the muddy sea-bed of the abyssal region, and some have penetrated into estuaries and harbours, where the water may be polluted and of reduced salinity. Many are fixed to firm substrates such as rock, stones, shells, the bodies of other animals or the surface of algae and others are loosely attached to submerged sand and mud, or partly embedded.

With a very few exceptions, none of which are British species, they are filter-feeding animals making use of the phytoplankton and organic detritus which they strain from a current of water drawn into the branchial sac through the oral siphon. The current is propelled by cilia lining the branchial stigmata and the food particles are trapped in mucus secreted by the endostyle. Mucus and entrapped food are passed across the branchial walls to the line of dorsal languets and drawn back into the oesophagus.

Sexual reproduction takes place in all ascidians. In temperate and cold seas breeding is generally restricted to the warmer season but in tropical waters it may continue throughout the year. Most species of solitary ascidian shed their eggs and sperm into the water where fertilization and development proceed. Cleavage is regular and the first division marks off the left and right halves of the future larva. Gastrulation is by invagination and the whole course of development from fertilization to the completion and hatching of the larva often occupies no more than one day.

Figure 2 illustrates the larva not of *Ciona* but of a compound form, since this shows certain structures not developed in *Ciona* and since it is only in compound ascidians that larvae are commonly seen, during incubation. The tadpole larva is peculiar to ascidians and consists of a trunk and a slender muscular tail, both covered with test, which is further developed as a flattened fin around the tail. At its anterior end the trunk bears a set of *papillae*, usually three in number, which are adhesive organs used when the larva settles before metamorphosis. The trunk houses the anterior part of the central nervous system in the form of a *cerebral vesicle*, containing the light-sensitive *ocellus* and the gravity-sensitive *otolith*. A few ascidians lack one or other of these sensory organs and in some they are combined as a *photolith* of dual function. The remaining part of the central nervous system extends within the tail, dorsal to the *notochord*. Also in the trunk is the precociously developed rudiment of the adult body; the degree to which this is developed varies considerably throughout the group.

The larva does not feed and its functions are dispersal and the choice of site for settlement. It may swim for only a few minutes or for many hours before fixing and metamorphosing to the adult form. During the early part of its pelagic life the larva tends to swim upwards but later this behaviour is reversed. Little is known regarding the mechanism of site choice but it is evidently important in relation to the habitat of the adult.

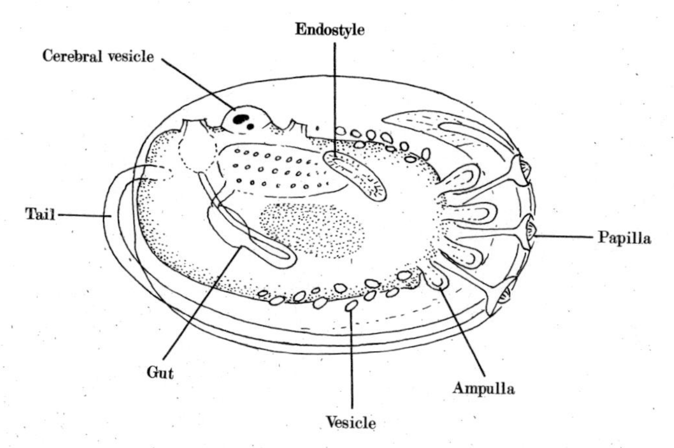

Fig. 2. Larva of a polyclinid ascidian, from the right, with the tail curled around the trunk (as seen within the egg membrane when zooids are dissected).

Asexual reproduction occurs in several families, by budding, and the resulting zooids remain in association to constitute a colony. Details of the process vary considerably, as do the tissues and organs involved. The colonies of many compound ascidians, particularly amongst the Polyclinidae and Clavelinidae, undergo a process of dedifferentiation after the breeding season and the resulting red or orange cylindrical bodies are often seen on rocky shores in the winter.

There is little information regarding the enemies of ascidians, except that their predators include bottom-feeding fish, carnivorous gastropods and starfish. A number of parasites and many commensal organisms are recorded.

Most species which have been studied have a life-span of 12–20 months, although a few are believed to live for several years. Death may result from senescence, from the onset of adverse conditions or by the attacks of enemies.

Collection and Preservation

Littoral species can be collected most readily from the lower sides of stones on sheltered rocky shores. Sub-littoral species are taken by most kinds of dredge and trawl. Specimens for routine collections and identification are adequately preserved in 10% neutral formalin or in 70% alcohol, but it is important to avoid the use of acid which destroys the calcareous spicules present in the test of some compound forms. It is not generally necessary to narcotize specimens before preservation if identification only is required. A few species, including *Ciona intestinalis*, contract badly unless narcotized but animals left for several hours in a dish of sea-water containing crystals of menthol can be fixed in an expanded condition.

Classification

Order ENTEROGONA

Suborder Aplousobranchiata

Family Clavelinidae

Subfamily Clavelininae

Clavelina lepadiformis (Müller)
Pycnoclavella aurilucens Garstang

Subfamily Holozoinae

Distaplia rosea Della Valle

Subfamily Polycitorinae

Archidistoma aggregatum Garstang
Polycitor searli Kott

Family Polyclinidae

Polyclinum aurantium Milne-Edwards
Synoicum pulmonaria (Ellis and Solander)
Morchellium argus (Milne-Edwards)
Sidnyum turbinatum Savigny
Sidnyum elegans (Giard)
Aplidium pallidum (Verrill)
Aplidium nordmanni (Milne-Edwards)
Aplidium proliferum (Milne-Edwards)
Aplidium punctum (Giard)
Aplidium glabrum (Verrill)

Family Didemnidae

Trididemnum tenerum (Verrill)
Didemnum candidum Savigny
Didemnum helgolandicum Michaelsen
Didemnum gelatinosum Milne-Edwards
Leptoclinides faeroensis Bjerkan
Diplosoma listerianum (Milne-Edwards)
Lissoclinum argyllense Millar

Suborder Phlebobranchiata

Family Cionidae

Ciona intestinalis (Linnaeus)

Family Diazonidae

Diazona violacea Savigny

Family Perophoridae

Perophora listeri Forbes

Family Corellidae
Corella parallelogramma (Müller)

Family Ascidiidae
Ascidiella aspersa (Müller)
Ascidiella scabra (Müller)
Ascidia mentula Müller
Ascidia conchilega Müller
Ascidia virginea Müller
Ascidia obliqua Alder
Ascidia prunum Müller
Phallusia mammillata (Cuvier)

Order PLEUROGONA

Suborder Stolidobranchiata

Family Styelidae
Pelonaia corrugata Forbes and Goodsir
Styela coriacea (Alder and Hancock)
Styela partita (Stimpson)
Styela clava Herdman
Cnemidocarpa mollis (Stimpson)
Polycarpa pomaria (Savigny)
Polycarpa gracilis Heller
Polycarpa rustica (Linnaeus)
Polycarpa fibrosa (Stimpson)
Dendrodoa grossularia (Van Beneden)
Distomus variolosus Gaertner
Stolonica socialis Hartmeyer
Botryllus schlosseri (Pallas)
Botrylloides leachi (Savigny)
Protostyela heterobranchia Millar

Family Pyuridae
Boltenia echinata (Linnaeus)
Microcosmus claudicans (Savigny)
Pyura tesselata (Forbes)
Pyura squamulosa (Alder)
Pyura microcosmus (Savigny)

Family Molgulidae
Molgula manhattensis (De Kay)
Molgula oculata Forbes
Molgula occulta Kupffer
Molgula citrina Alder and Hancock
Molgula complanata Alder and Hancock
Eugyra arenosa (Alder and Hancock)

Key to the Families of British Ascidiacea

Note: The key applies to families as represented by British species but not necessarily to species from other areas.

1. Solitary or colonial; body undivided or divided into 2 or 3 regions; branchial sac without folds; gonads unpaired, in or below gut loop (Order Enterogona) **2**

Solitary or colonial; body undivided; branchial sac with or without folds; gonads in body wall (Order Pleurogona) **9**

2. Colonial; body divided into thorax and abdomen and sometimes post-abdomen; branchial sac without internal longitudinal bars; gut loop in abdomen; gonads in abdomen or post-abdomen (Suborder Aplousobranchiata) **3**

Solitary or colonial; body undivided (except in Diazonidae); branchial sac with internal longitudinal bars (sometimes incomplete) (Suborder Phlebobranchiata) **5**

3. Zooids embedded; colonies form thin encrusting sheets; common test usually has calcareous spicules; body divided into thorax and abdomen; 3 or 4 rows of branchial stigmata; gonads beside gut loop; larvae in common test below zooids . . . . . Family Didemnidae (p. 32)

Zooids embedded or free; common test without spicules; body divided into thorax, abdomen and sometimes post-abdomen; larvae in atrial cavity of zooid **4**

4. Zooids with thorax and abdomen . . . . Family Clavelinidae (p. 11)

Zooids with thorax, abdomen and post-abdomen, elongated and embedded in a fleshy common test . . . . . . Family Polyclinidae (p. 17)

5. Solitary; gut loop below and to right of branchial sac; no papillae on internal longitudinal branchial bars; stigmata spiral Family Corellidae (p. 42)

Solitary or colonial; gut loop not to right of branchial sac; internal longitudinal branchial bars with or without papillae; stigmata straight . **6**

6. Colonial; zooids free but united by basal stolons; eggs incubated to larval stage . . . . Family Perophoridae (p. 41)

Solitary or, when colonial, with zooids embedded; eggs not incubated **7**

7. Solitary or colonial; body divided into thorax and abdomen; internal longitudinal branchial bars, when present, without papillae; gut loop below branchial sac, in abdomen . . . Family Diazonidae (p. 40)

Solitary; body not divided; internal longitudinal branchial bars with papillae (except *Ascidiella*); gut loop below or to left of branchial sac . . **8**

8. Test soft; gut loop below branchial sac; dorsal languets present; heart V-shaped . . . . Family Cionidae (p. 39)

Test firm; gut loop to left of branchial sac; continuous dorsal lamina; heart straight . . . . Family Ascidiidae (p. 43)

9. Solitary or colonial; siphons plain or with 4 lobes; oral tentacles simple; not more than 4 branchial folds; stigmata straight; dorsal lamina continuous; stomach without hepatic diverticula Family Styelidae (p. 53)

Solitary; siphons with 4 or 6 lobes; oral tentacles usually compound; branchial folds, when present, often more than 4; dorsal lamina continuous or replaced by dorsal languets; stigmata straight or spiral, stomach with hepatic diverticula **10**

10. Siphons with 4 lobes; stigmata straight; no large renal sac Family Pyuridae (p. 73)

Siphons with 6 lobes; stigmata spiral; one large renal sac on right side Family Molgulidae (p. 78)

Systematic Part

In the description of species there is given, below the name of the species, the name under which it was first described and, in some cases, a few of the synonyms.

Family CLAVELINIDAE

1. Each zooid free, except for basal stolons or continuous mass of basal test; siphons without lobes **2**

Zooids embedded or in free-standing groups each of a few zooids united in common test; siphons 6-lobed **3**

2. Zooids joined by stolons or thin basal test; zooids up to 2 cm long, with up to 17 rows of stigmata. . . . . . *Clavelina lepadiformis* (p. 12)

Zooids joined by thick mass of basal test; zooids up to 6 mm long, with up to 9 rows of stigmata . . . . . . *Pycnoclavella aurilucens* (p. 13)

3. Atrial siphons open directly on to surface of colony **4**

Atrial siphons open into common cloacal cavities, which have common cloacal openings on surface of colony . . . *Distaplia rosea* (p. 14)

4. Zooids with 3 rows of stigmata . . . *Archidistoma aggregatum* (p. 15)

Zooids with 4 rows of stigmata . . . . . . *Polycitor searli* (p. 16)

Genus CLAVELINA Savigny, 1816

Clavelina lepadiformis (Müller, 1776)

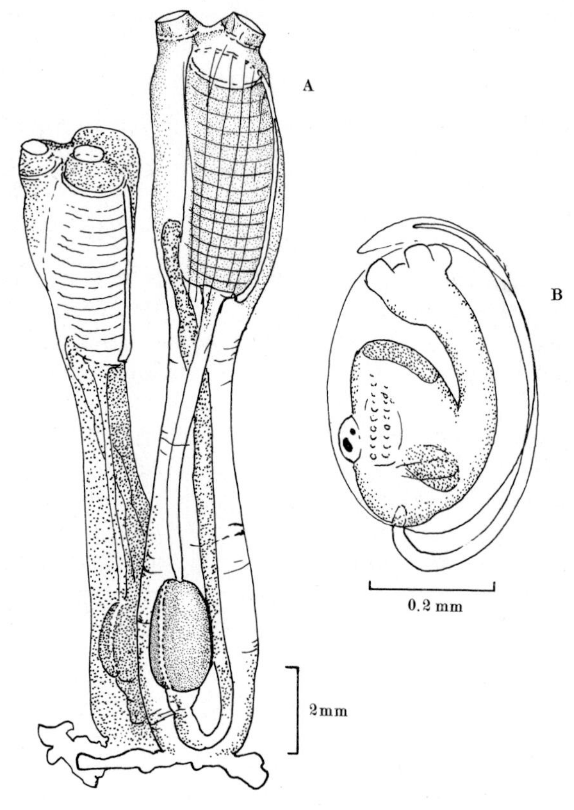

Fig. 3. *Clavelina lepadiformis*: **A,** zooids; **B,** larva.

Ascidia lepadiformis Müller, 1776

Colony consists of a bunch of zooids almost completely free; zooids up to 2 cm long, transparent and delicate, with white, yellow or pink markings delineating the endostyle and other ciliated tracts of the branchial sac; thorax with up to 17 rows of stigmata; oesophagus and rectum long and narrow; gonads beside lower part of intestinal loop; eggs and larvae free in atrial cavity of zooid; larval trunk 0·4–0·5 mm long, with ventral process bearing 3 papillae in triangular arrangement.

From low-water level to about 50 m, on seaweed, shells, stones, etc. Generally distributed but local in British waters, and elsewhere from western Norway to the Adriatic.

Genus PYCNOCLAVELLA Garstang, 1891

Pycnoclavella aurilucens Garstang, 1891

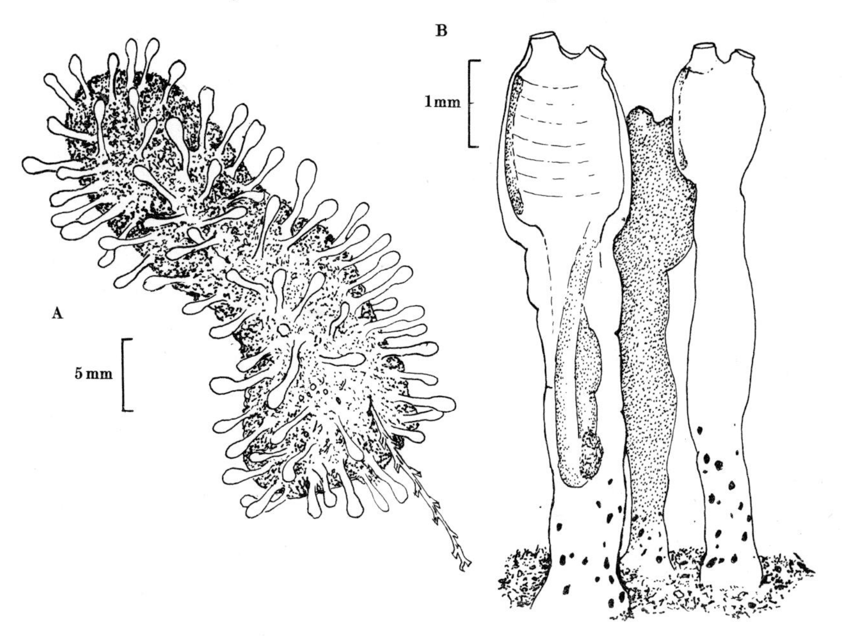

Fig. 4. *Pycnoclavella aurilucens*: **A**, colony; **B**, zooids.

Pycnoclavella aurilucens Garstang, 1891

Colony consists of a thick basal mass of common test from which project the small finger-shaped zooids, each up to about 6 mm in length. Living specimens, according to Berrill (1950), have bright pigment on the thorax, as in *Clavelina*; thorax short, with up to 9 rows of stigmata; gut and gonads much as in *Clavelina*; eggs and larvae incubated in terminal part of oviduct.

In British waters recorded only from 25–30 m near Plymouth.

Genus Distaplia Della Valle, 1881

Distaplia rosea Della Valle, 1881

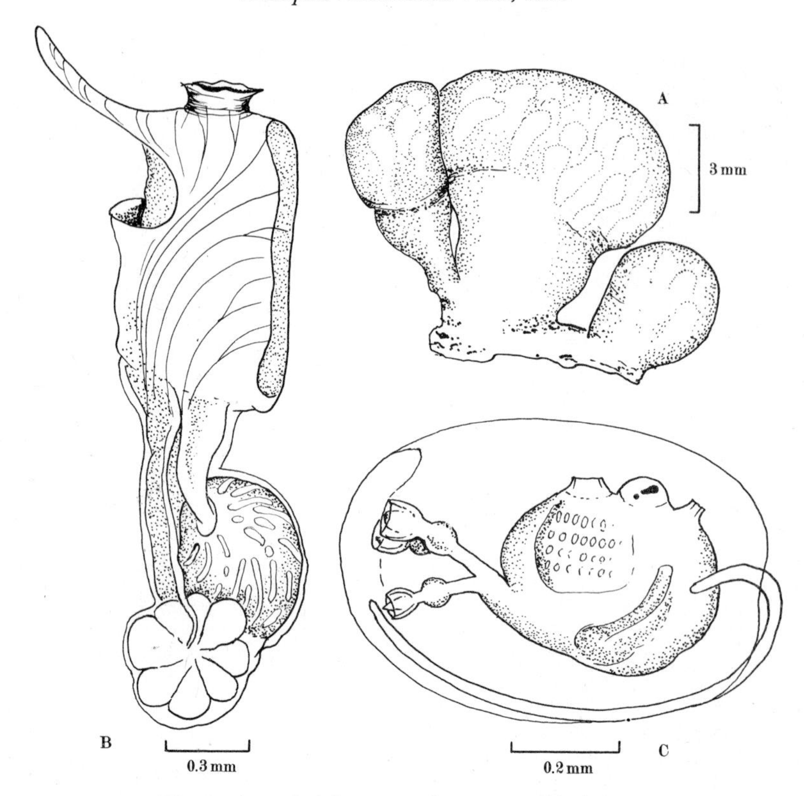

Fig. 5. *Distaplia rosea*: **A,** colony; **B,** zooid; **C,** larva.

Distaplia rosea Della Valle, 1881
Distaplia garstangi Berrill, 1950

Colony of a few soft pink dome-shaped or nearly globular masses either with a short stalk or sessile and united by a basal plate of test; these masses may reach 1 cm or more in diameter; zooids with conspicuous atrial languet, 4 rows of stigmata, stomach with numerous short folds and testis forming a rosette of follicles below the stomach. Larvae, which develop in a brood sac projecting from the dorsal part of the thorax, have 2 dorsal and one ventral papillae arising from a ventral process of the trunk.

Known from the English Channel and the coasts of Wales, Norfolk and south-west Scotland. It is also recorded from the Mediterranean.

Berrill (1950) has described a species *D. garstangi* from Plymouth but it appears to differ little from *D. rosea* and is here regarded as a synonym.

Archidistoma aggregatum Garstang, 1891

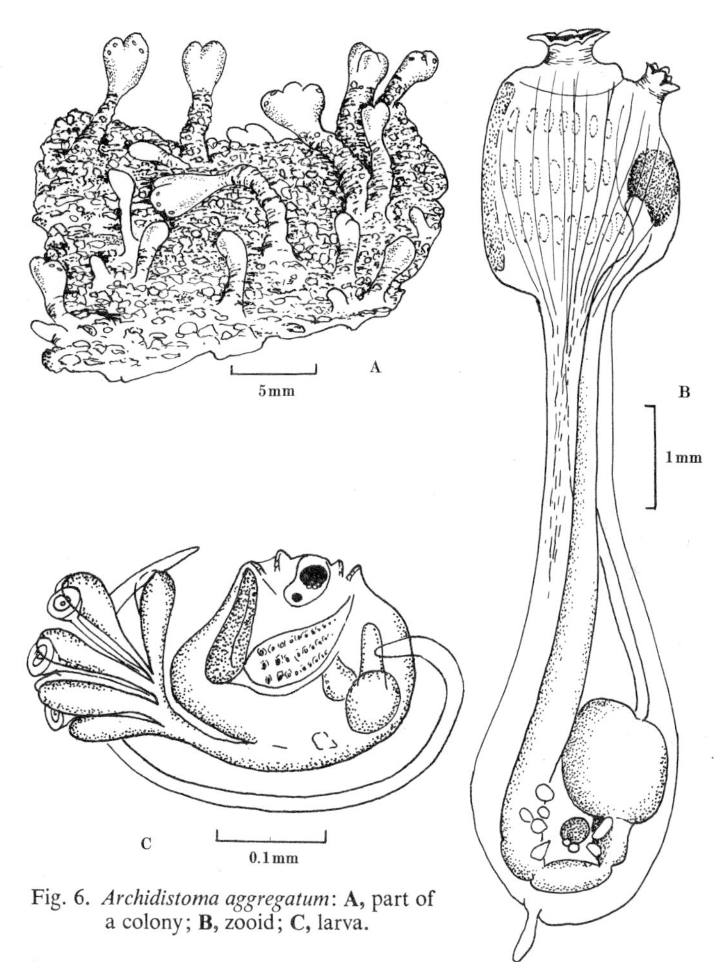

Fig. 6. *Archidistoma aggregatum*: **A,** part of a colony; **B,** zooid; **C,** larva.

Archidistoma aggregatum Garstang, 1891

Colony inconspicuous, consisting of a thin mass of common test with finger-like projections about 1 cm long; each projection is thickened and partially divided at the end and accommodates a few zooids; zooids with 6-lobed oral and atrial siphons, a short thorax with 3 rows of stigmata and a long abdomen; oesophagus and rectum long and narrow; stomach globular and smooth-walled; gonads beside lower intestinal loop. Larvae, in atrial cavity of zooid, with 3 papillae in a vertical row.

Recorded from depths of 0–30 m near Plymouth. It may occur on the French coast and has been reported from North Carolina, U.S.A.

Genus Polycitor Renier, 1804

Polycitor searli Kott, 1952

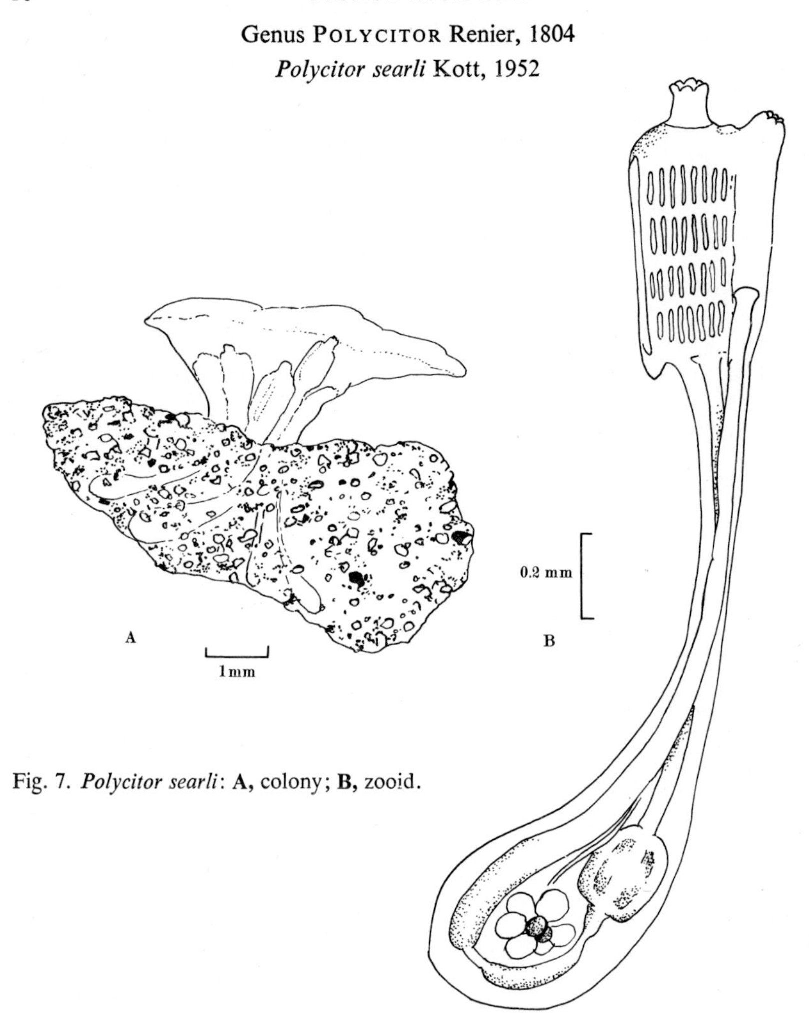

Fig. 7. *Polycitor searli*: **A,** colony; **B,** zooid.

Polycitor searli Kott, 1952

Colony inconspicuous, up to 6 mm high, and consisting of an upper clear portion and a sand-coated base. The colony has 4–12 zooids; both siphons 6-lobed and both opening on surface of colony; 4 rows of stigmata; oesophagus and rectum long and narrow; stomach wall smooth externally but with 4 folds internally; gonads in lower intestinal loop. Larvae, which are incubated, have 3 papillae in a vertical row.

The species is known only from the original record by Kott (1952), from the neighbourhood of Plymouth.

Family POLYCLINIDAE

1. Stomach smooth; lower gut loop twisted; post-abdomen joined to abdomen by a narrow stalk . . . . *Polyclinum aurantium* (p. 18)

Stomach with folds or rounded swellings; lower gut loop not twisted; post-abdomen without distinct narrow stalk 2

2. Stomach with numerous rounded swellings, or markings 3

Stomach with longitudinal folds, which may be broken into short lengths 4

3. Oral siphon with 6 lobes . . . . *Synoicum pulmonaria* (p. 20)

Oral siphon with 8 lobes . . . . *Morchellium argus* (p. 22)

4. Oral siphon with 8 lobes 5

Oral siphon with 6 lobes 6

5. Colony usually with numerous small narrow-based lobes; 7–9 rows of stigmata; 10–12 folds on stomach . . *Sidnyum turbinatum* (p. 24)

Colony usually with a few broad lobes; 11–15 rows of stigmata; up to 20 folds on stomach . . . . *Sidnyum elegans* (p. 26)

6. Atrial opening of zooids without languet and distinctly posterior to oral siphon; post-abdomen short . . . . . *Aplidium pallidum* (p. 27)

Atrial opening with languet and at anterior end of thorax; post-abdomen long when fully developed 7

7. Stomach with 6 folds . . . . *Aplidium punctum* (p. 30)

Stomach with more than 6 folds 8

8. Stomach with 10–20 folds . . . . . . *Aplidium glabrum* (p. 31)

Stomach with 20–30 folds 9

9. Colony of flat-topped, squat and sessile lobes; systems of zooids distinct and regular . . . . *Aplidium nordmanni* (p. 28)

Colony of ovoid to club-shaped lobes with rounded top; systems of zooids indistinct . . . . *Aplidium proliferum* (p. 29)

Genus POLYCLINUM Savigny, 1816

Polyclinum aurantium Milne-Edwards, 1841

Polyclinum aurantium Milne-Edwards, 1841
Polyclinum cerebriforme Alder and Hancock, 1912

Colony consists of groups of rounded, globular or flat-topped heads up to 3 cm in diameter; dull yellow and usually coated with sand; each head with one to a few common cloacal openings. Zooid with large thorax, smaller abdomen, and post-abdomen joined to abdomen by a slender stalk. Branchial sac with 10–16 rows of stigmata and small papillae on transverse bars; stomach smooth-walled and ovoid; lower part of gut loop twisted; gonad in post-abdomen and consisting of small ovary embedded in mass of male follicles. Larvae in atrial cavity; trunk about 0·5 mm long, with 3 papillae in vertical row and 4 pairs of lateral ampullae.

On rock, etc., from lower shore to about 100 m; generally distributed round British coasts and elsewhere from western Norway to the Mediterranean.

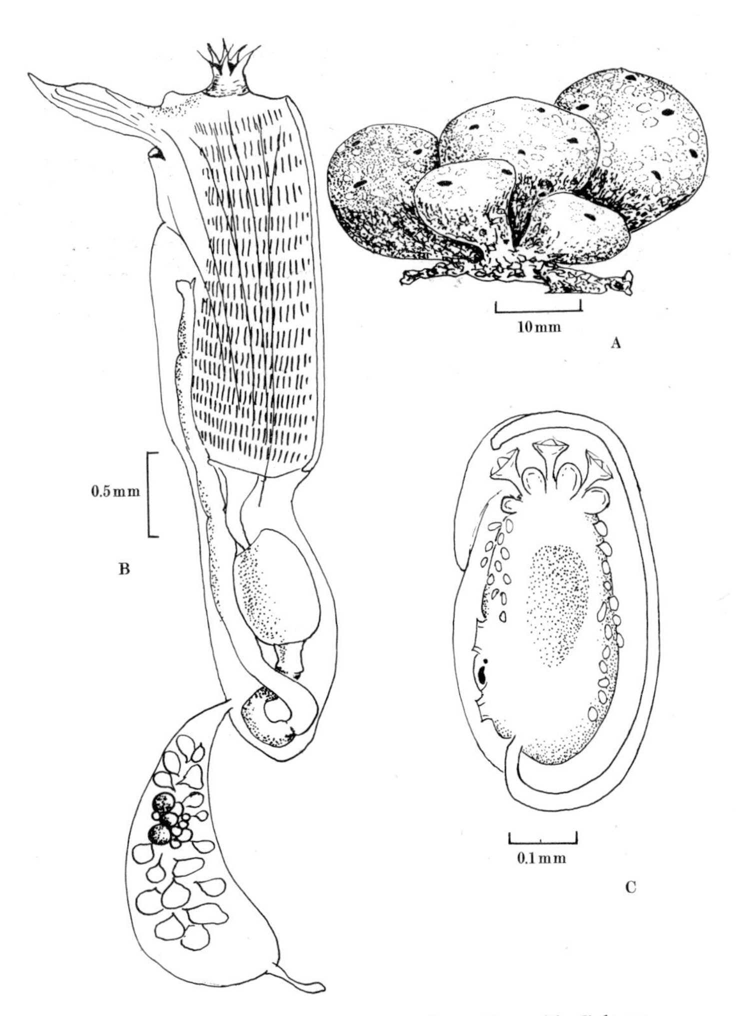

Fig. 8. *Polyclinum aurantium*: **A,** colony; **B,** zoid; **C,** larva.

Genus SYNOICUM Phipps, 1774

Synoicum pulmonaria (Ellis and Solander, 1786)

Alcyonium pulmonaria Ellis and Solander, 1786
Amaroucium pomum and *Aplidium ficus* Alder and Hancock, 1912

Colonies up to 10 cm in diameter and globular or consisting of groups of smaller ovoid heads; semi-transparent and smooth or with adhering sand; systems of zooids numerous, round and regular, each with a central common cloacal opening. Zooids with a 3-lobed atrial languet and short tubular atrial siphon; 8–20 rows of stigmata; stomach with numerous rounded markings or low swellings; ovary below or extending forward beside lower intestinal loop; testis behind ovary and may extend forward beside it. Larvae in atrial cavity; trunk 0·6–1·0 mm long, with 3 papillae in vertical line and with median and paired lateral ampullae.

On stones, shell, etc., from shallow water to 600 m. Recorded off the North Sea coasts of the British Isles and also sparingly off the south and west coasts. The species is distributed from the Arctic to the English Channel and off the north-east coast of America.

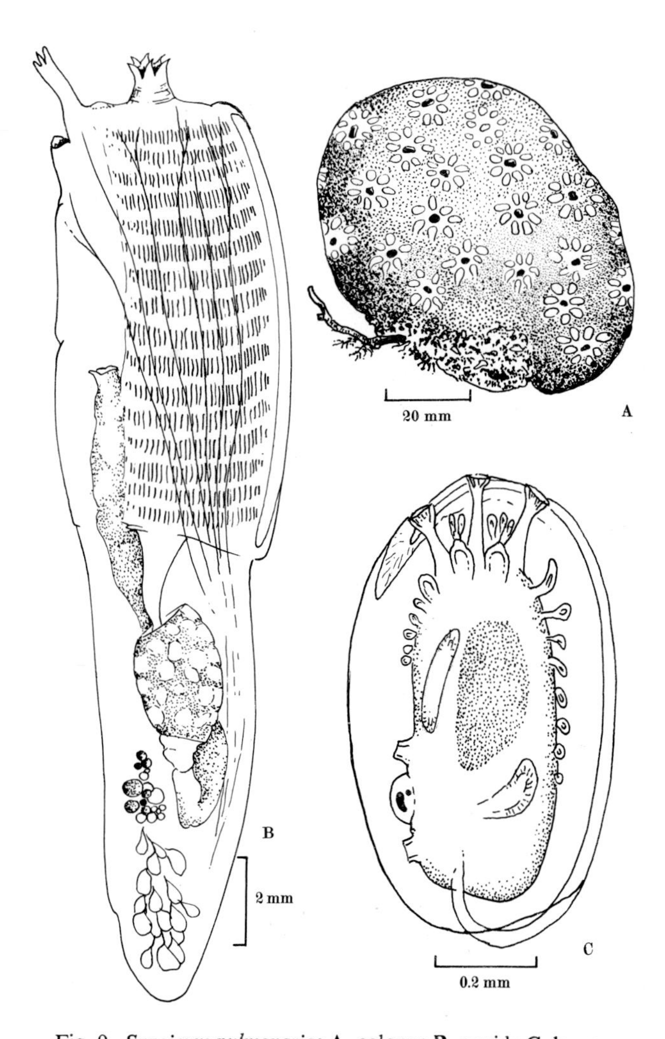

Fig. 9. *Synoicum pulmonaria*: **A,** colony; **B,** zooid; **C,** larva.

Genus Morchellium Giard, 1872

Morchellium argus (Milne-Edwards, 1841)

Amaroucium argus Milne-Edwards, 1841

Colony of pink or red lobes joined at base, each lobe having a long firm sand-coated stalk and a wider rounded bare head; zooids visible on heads. Zooid with 8 lobes on oral siphon and small pointed atrial languet; 4 red spots at base of oral siphon; 12–20 rows of stigmata; stomach with numerous rounded swellings; ovary below intestinal loop; testis elongate, behind ovary. Larvae in atrial cavity; trunk about 0·8 mm long, with 3 papillae in vertical line and a fringe of small vesicles around anterior part.

From the lower shore to shallow water, on the south and west coasts of Britain northwards to south-west Scotland and elsewhere known from the French coast.

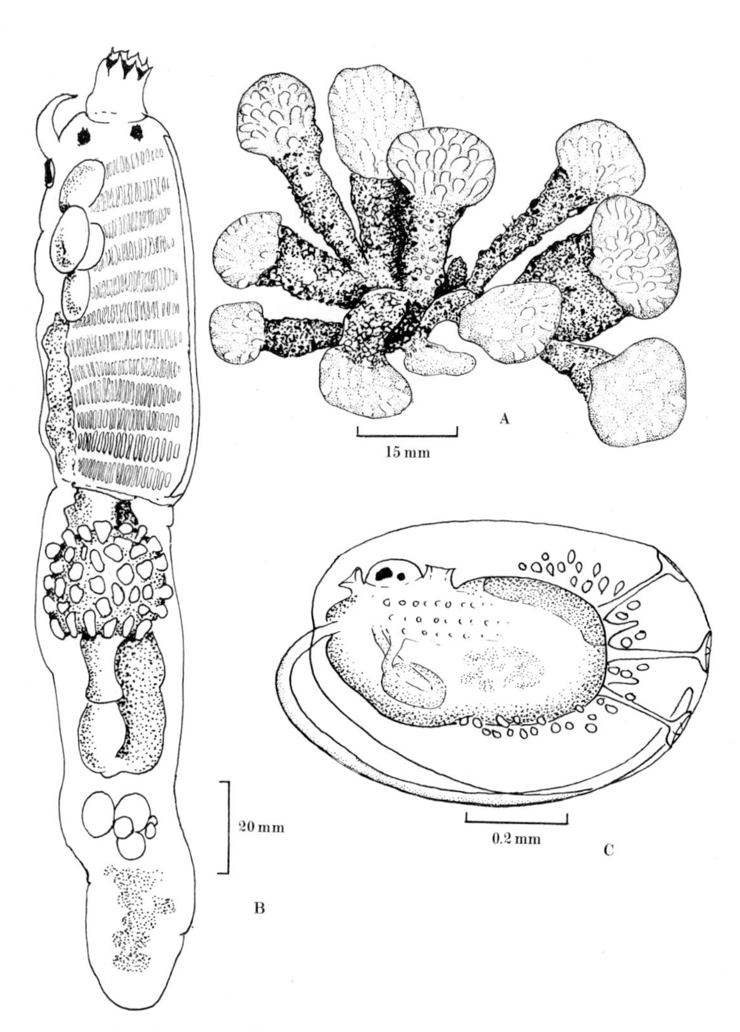

Fig. 10. *Morchellium argus*: **A,** colony; **B,** zooid; **C,** larva

Genus SIDNYUM Savigny, 1816

Sidnyum turbinatum Savigny, 1816

Sidnyum turbinatum Savigny, 1816
Polyclinum succineum, Parascidia forbesii and *P. flemingi* Alder and Hancock, 1912

Colonies consist of a few club-shaped, ovoid and usually rather flat-topped heads, some of which may be broad and squat; heads united at their lower narrow ends by creeping stolons; each head with a common cloacal opening; test semi-transparent and without encrusting sand; colony 1–2 cm high. Zooids with 8 oral lobes, a pair of red-orange pigment spots at base of oral siphon, 7–9 rows of stigmata, 10–15 folds (often broken into short lengths) on the stomach; post-abdomen often short, ovary below intestinal loop and male follicles below ovary. Larvae in atrial cavity; trunk 0·4–0·6 mm long, 3 papillae in a vertical line; vesicles arranged around anterior part of trunk.

From the lower shore down to 200 m, on rock, stones, etc. Generally distributed around British Isles, mainly in sheltered rocky parts of the south, west and north coasts. It occurs from western Norway to the Mediterranean.

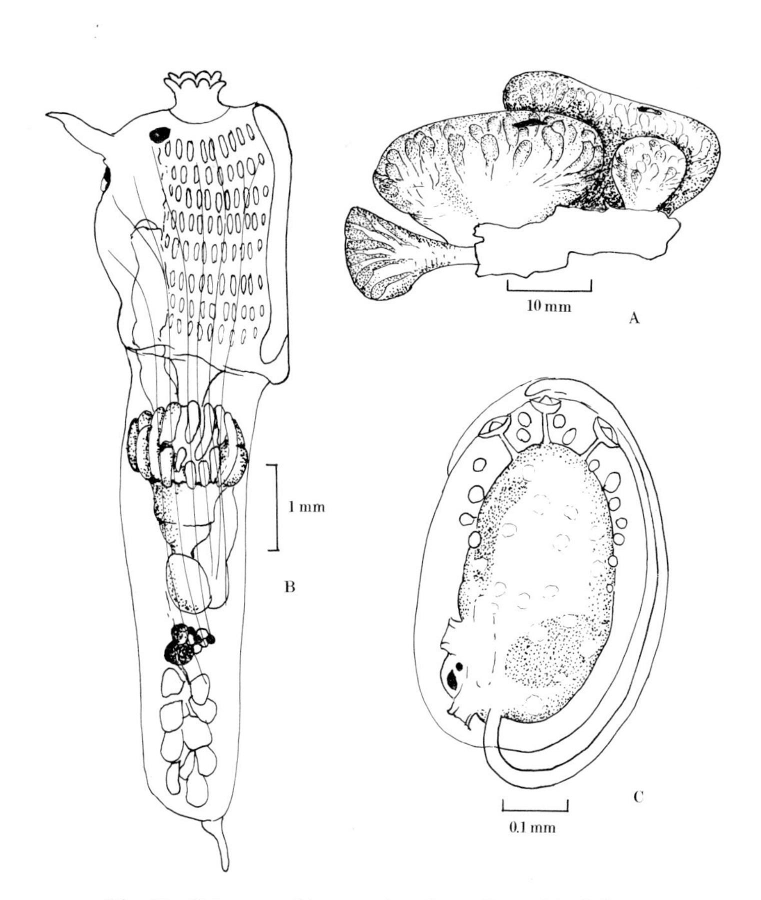

Fig. 11. *Sidnyum turbinatum*: **A,** colony; **B,** zooid; **C,** larva.

Aplidium nordmanni (Milne-Edwards, 1841)

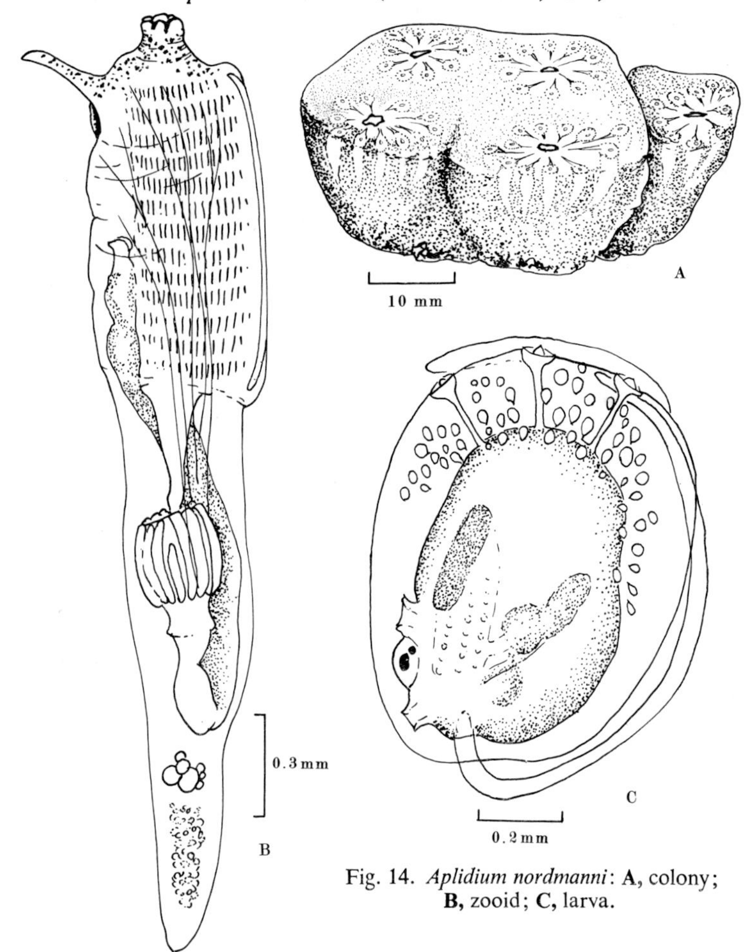

Fig. 14. *Aplidium nordmanni*: **A,** colony; **B,** zooid; **C,** larva.

Amaroucium nordmanni Milne-Edwards, 1841

Colonies simple or partially divided into a few closely adhering lobes; flat-topped squat and broadly based; large specimens 4–6 cm in diameter; pink and white with a pattern of zooids arranged in systems and central common cloacal openings. Zooids with simple pointed atrial languet, 9–13 rows of stigmata, 20–30 folds on stomach (some folds short or broken), post-abdomen having anterior ovary and posterior biserially arranged male follicles. Larvae in atrial cavity; trunk about 0·9 mm long; 3 papillae in vertical row; numerous small vesicles around anterior end.

On lower shore and shallow water, on stones, rock, etc. South and west coasts of British Isles and elsewhere south to the Mediterranean.

Aplidium proliferum (Milne-Edwards, 1841)

Amaroucium proliferum Milne-Edwards, 1841
Amaroucium albicans and *A. papillosum* Alder and Hancock, 1912

It is not certain that this species should be recognized as distinct from *A. nordmanni*; Berrill (1950) retains both species but Kott (1952) considers them forms of one species, *A. proliferum*. If the distinction is maintained, it depends on the stalked club-shaped lobes of the colony of *A. proliferum* in contrast to the flat-topped sessile colonies of *A. nordmanni*.

In view of the confusion between *A. proliferum* and *A. nordmanni*, it is impossible to be sure of the distribution but the former species may occur locally around all British coasts and in the Mediterranean, on the lower shore and down to 50 m.

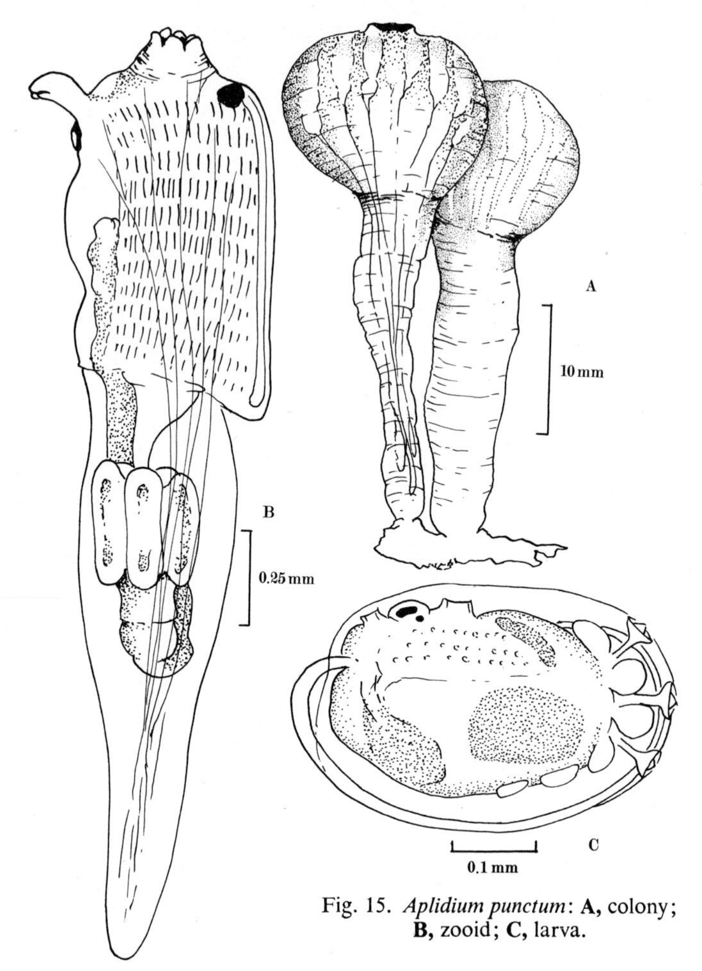

Fig. 15. *Aplidium punctum*: **A,** colony; **B,** zooid; **C,** larva.

Amaroucium punctum Giard, 1873
Distoma vitreum Alder and Hancock, 1912

Colonies of slender club-shaped or conical lobes attached by a narrow base; clear and delicate in appearance; up to 4 cm long. Zooids with 3-lobed atrial languet, 9–12 rows of stigmata, a conspicuous red spot over the anterior end of the endostyle; stomach with 6 folds; post-abdomen with anterior ovary and posterior biserially arranged male follicles. Larvae in atrial cavity; trunk 0·4–0·5 mm long; 3 papillae in vertical row; ampullae but no vesicles around anterior end.

On stones and rocks from lower shore to shallow water. South and west coasts of British Isles and in English Channel.

Aplidium glabrum (Verrill, 1871)

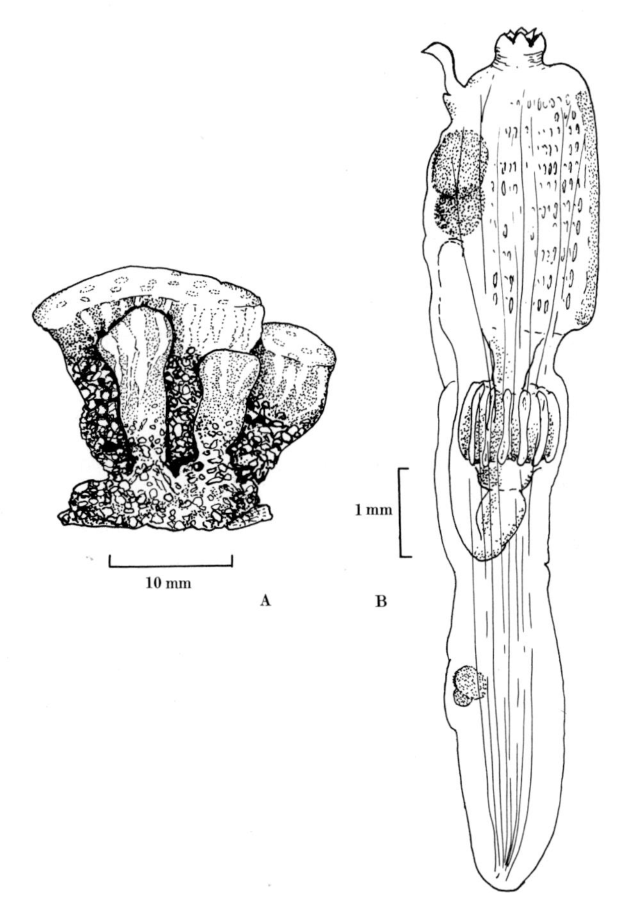

Fig. 16. *Aplidium glabrum*: **A,** colony; **B,** zooid.

Amouroucium glabrum Verrill, 1871

Colonies of several upright, usually flat-topped lobes, with bare upper surface and sand-coated sides and base; test grey and semi-transparent; up to 2 cm in height. Zooids with simple or slightly 3-lobed atrial languet, 10–12 rows of stigmata, 10–20 folds on stomach, post-abdomen with anterior ovary and posterior biserial male follicles. Larva unknown.

On shell, stones, etc. from low water line to 400 m. North-west and northern Scottish waters and elsewhere extends to the Arctic.

Family DIDEMNIDAE

1. Zooids with 3 rows of stigmata . . . . *Trididemnum tenerum* (p. 33)
Zooids with 4 rows of stigmata **2**

2. Atrial opening simple, without siphon **3**
Backwardly directed tubular atrial siphon *Leptoclinides faeroensis* (p. 36)

3. Atrial opening large, exposing most of branchial walls; sperm duct not spirally coiled **4**
Atrial opening often small or of moderate size (although sometimes exposing much of branchial walls); sperm duct spirally coiled **5**

4. No calcareous spicules in common test* . *Diplosoma listerianum* (p. 37)
Calcareous spicules in common test . . *Lissoclinum argyllense* (p. 38)

5. Common test with a few spicules grouped around zooids, or without spicules *Didemnum gelatinosum* (p. 35)
Common test with many, closely packed spicules **6**

6. Larva with 2 papillae . . . . *Didemnum candidum* (p. 34)
Larva with 3 papillae . . . . . *Didemnum helgolandicum* (p. 35)

* Carlisle (1953) reported minute spicules in living specimens.

Genus TRIDIDEMNUM Della Valle, 1881

Trididemnum tenerum (Verrill, 1871)

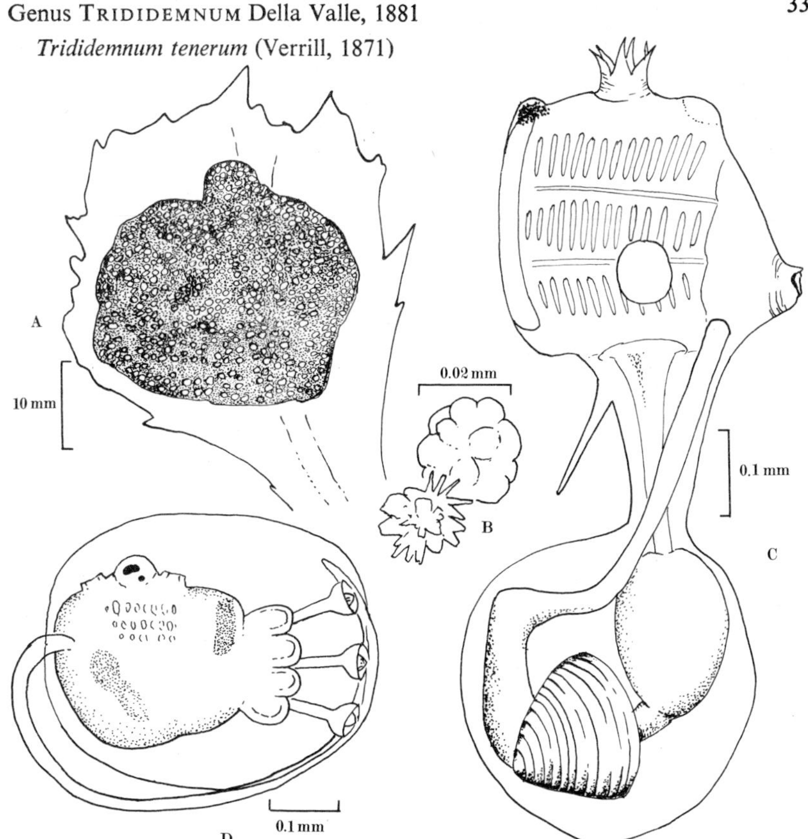

Fig. 17. *Trididemnum tenerum*: **A,** colony; **B,** spicules; **C,** zooid; **D,** larva. The colony, which is encrusting a frond of *Fucus serratus*, is shown darker than most appear in life.

Lissoclinum tenerum Verrill, 1871

Colonies encrusting, a few mm thick, white, grey or semi-transparent and gelatinous; spicules vary from a few to abundant and from irregular groups of rods to solid stellate bodies. Zooids with 3 rows of stigmata, a tubular atrial siphon and usually a dark pigment spot over the anterior end of the endostyle; testis undivided; sperm duct usually with 10–12 spiral turns. Larva with 3 papillae in vertical row and 4 pairs of anterior ampullae; trunk 0·4–0·6 mm long.

On algae, stones, etc. from lower shore to about 200 m. On the west and south coasts of British Isles and extending northwards to Arctic seas.

Carlisle (1953) has recorded another species, *T. niveum* (Giard) from the Plymouth area, but the distinctions between this species and *T. tenerum* appear to be slight and unreliable when single or few specimens are available.

Trididemnum alleni Berrill, 1947 was described from the Plymouth area but Carlisle (1954) has shown it to represent small specimens of *Didemnum candidum*.

Genus DIDEMNUM Savigny, 1816

Didemnum candidum Savigny, 1816

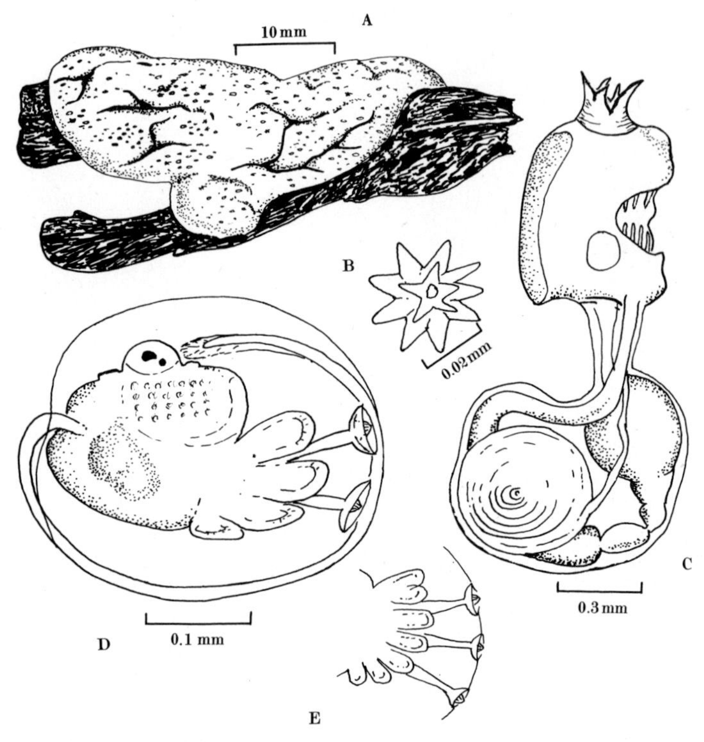

Fig. 18. *Didemnum candidum*: **A,** colony attached to a frond of *Fucus*; **B,** spicule; **C,** zooid; **D,** larva; **E,** anterior end of larva of *Didemnum helgolandicum*.

Didemnum candidum Savigny, 1816
Didemnum maculosum Berrill, 1950

Colonies hard and encrusting; white, grey, yellow or violet or marked with dark lines; spicules stellate, usually abundant. Zooids with simple atrial opening, 4 rows of stigmata, a narrow waist, testis undivided and sperm duct with 7–10 spiral turns. Larva with 2 papillae in vertical row and 4 pairs of anterior ampullae; trunk about 0·35 mm long.

On stones, rock, algae, etc. from the lower shore to moderately deep water. West and south coasts of British Isles and elsewhere through the Mediterranean and Red Sea and possibly other warm regions.

Didemnum helgolandicum Michaelsen, 1921

Didemnum helgolandicum Michaelsen, 1921

This species cannot be distinguished reliably from *D. candidum* by any characters of the colony, spicules or zooids (Carlisle, 1954). The larva of *D. helgolandicum*, however, has 3 papillae in contrast with that of *D. candidum* with 2 papillae. Both species have been recorded as *D. maculosum* (Milne-Edwards), and Carlisle (1954) has discussed the confused situation with regard to the three names.

The habitat is similar to that of *D. candidum* but *D. helgolandicum* has been recorded in British waters only from Plymouth; elsewhere from Heligoland, the Skagerrak, the Faeroes and west Norway.

Didemnum gelatinosum Milne-Edwards, 1841

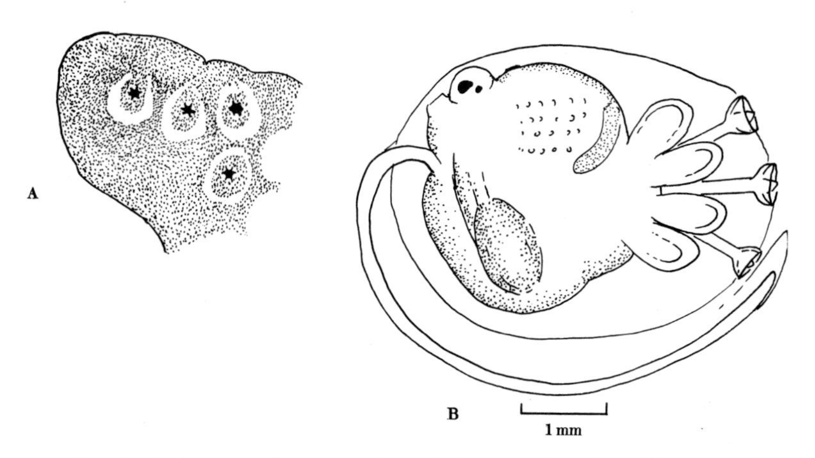

Fig. 19. *Didemnum gelatinosum*: **A,** part of colony, showing white zones of spicules; **B,** larva.

Didemnum gelatinosum Milne-Edwards, 1841

Colonies encrusting, transparent and sometimes with circular or crescentic white markings around the zooids; spicules absent or few and grouped around zooids. Zooids, spicules and larvae similar to those of *D. helgolandicum.* The species is characterized mainly by the scarcity of its spicules and the consequent gelatinous appearance of the common test. I am not sure that the British records relate to the species described by Milne-Edwards; they may even represent colonies of *D. helgolandicum.*

Recorded from the shore to depths of 50 m and in British waters only from Plymouth; elsewhere, in the Mediterranean.

Genus Leptoclinides Bjerkan, 1905

Leptoclinides faeroensis Bjerkan, 1905

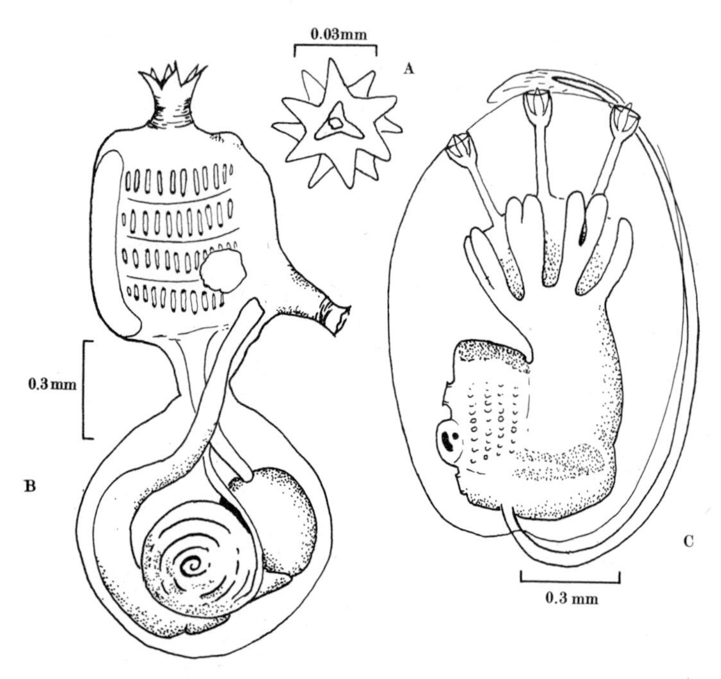

Fig. 20. *Leptoclinides faeroensis*: **A,** spicule; **B,** zooid; **C,** larva.

Leptoclinides faeroensis Bjerkan, 1905

Colonies encrusting, grey or white; spicules mainly in upper part of common test, stellate, up to 50 μ in diameter. Zooids with 4 rows of stigmata, backwardly directed tubular atrial siphon, undivided testis and 4 to 6 spiral turns of sperm duct. Larva with 3 papillae in vertical row, 4 pairs of lateral ampullae (sometimes bifid), trunk about 1·2 mm long.

On shells, stones and other hard objects, from the intertidal zone (record from Plymouth) to depths of 2000 m. In British waters recorded only from Plymouth; elsewhere, off the western Norwegian coast and in more northerly waters.

Genus Diplosoma MacDonald, 1859

Diplosoma listerianum (Milne-Edwards, 1841)

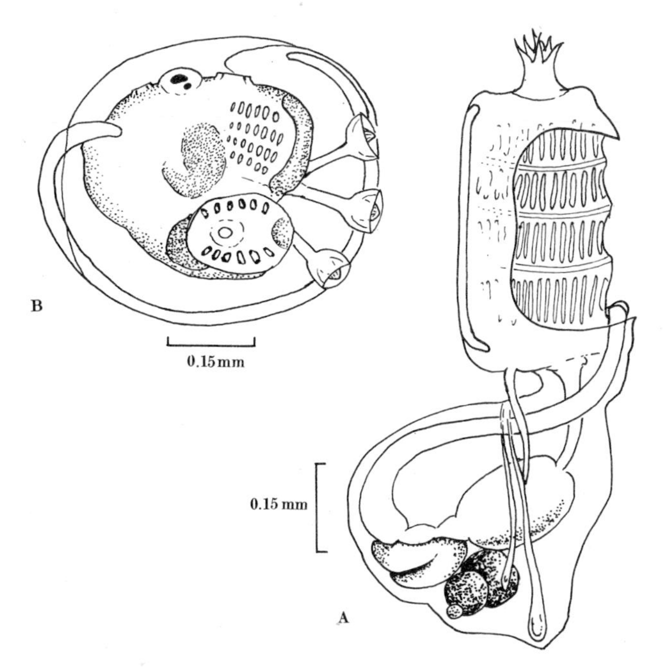

Fig. 21. *Diplosoma listerianum*: **A,** zooid; **B,** larva.

Leptoclinum listerianum Milne-Edwards, 1841
Leptoclinum gelatinosum, and *L. punctatum* Alder and Hancock, 1912

Colonies thin and sheeting, soft and gelatinous, semi-transparent and with dark spots; spicules absent but Carlisle (1953) noted minute spicules which he found to be readily destroyed in preservation of the colony. Zooids with wide atrial opening, 4 rows of stigmata, 2 closely united male follicles and sperm duct not spirally coiled but curved at its origin. Larvae with 3 papillae in vertical row and precocious bud consisting of thorax and abdomen below those of oozooid; trunk about 0·5 mm long.

On stones, piers, the test of solitary ascidians, etc., from low-water level to depths of about 80 m. Widely distributed in British waters, the west coast of Europe and the Mediterranean; particularly in sheltered areas.

Genus LISSOCLINUM Verrill, 1871

Lissoclinum argyllense Millar, 1950

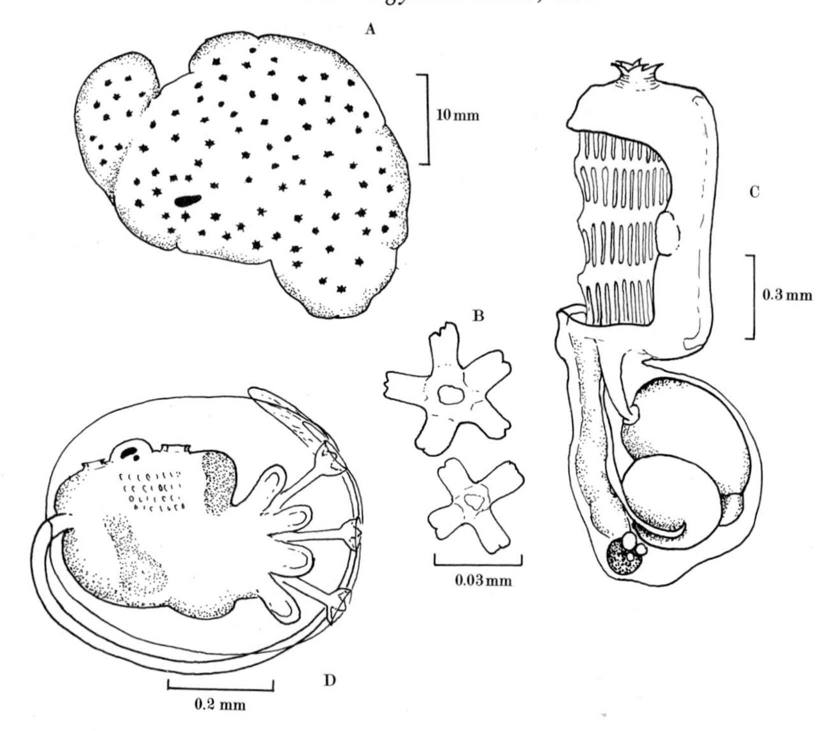

Fig. 22. *Lissoclinum argyllense*: **A,** colony; **B,** spicules; **C,** zooid; **D,** larva.

Lissoclinum argyllense Millar, 1950

Colonies white or creamy with small dark spots marking the oral openings and a few larger slit-like common cloacal openings; spicules densely packed, with 4–6 rays in optical section, the rays being blunt with a notched end. Zooid with large atrial opening, 4 rows of stigmata and single male follicle with uncoiled sperm duct curved at its origin. Larva with 3 papillae in vertical row and 4 pairs of anterior ampullae; trunk about 0·6 mm long.

On stones, etc. from the lower shore and shallow water. Recorded only from the Scottish west coast and from Plymouth.

Lissoclinum cupuliferum Kott, 1952 was described from a single colony found in 15 fathoms near Plymouth. Kott described this species as having small mulberry shaped spicules in the test but I have examined the original colony and found that the bodies identified as spicules are not dissolved in acid and, therefore, are not calcareous. They appear to be test-cells and the species, therefore, may be a *Diplosoma*. In many respects, including the larva with its precocious bud, *L. cupuliferum* resembles *D. listerianum* and, despite its tough test, it may represent that common British species.

Family CIONIDAE

Genus CIONA Fleming, 1822

Ciona intestinalis (Linnaeus, 1767)

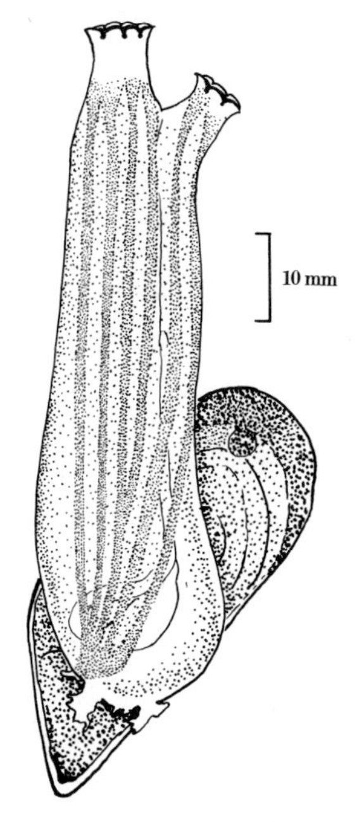

Fig. 23. *Ciona intestinalis*, attached to a mussel shell.

Ascidia intestinalis Linnaeus, 1767

Body cylindrical, soft and retractile, greenish and translucent, sometimes with orange bars on the body; longitudinal muscles conspicuous; oral siphon terminal, with 8 lobes and red or orange pigment spots between lobes; atrial siphon sub-terminal, with 6 lobes and pigment spots; branchial sac extending most of the length of body; stigmata straight; row of pointed dorsal languets; papillae on longitudinal branchial bars; stomach and intestinal loop behind branchial sac; ovary compact, in intestinal loop; testis diffusely spread over stomach and intestine; oviparous.

On stones, piers, algae, etc., from the lower shore to depths of at least 500 m. Generally distributed around British coasts and of very wide occurrence in many parts of the world.

Genus DIAZONA Garstang, 1891

Diazona violacea Savigny, 1816

A

20 mm

5 mm

B

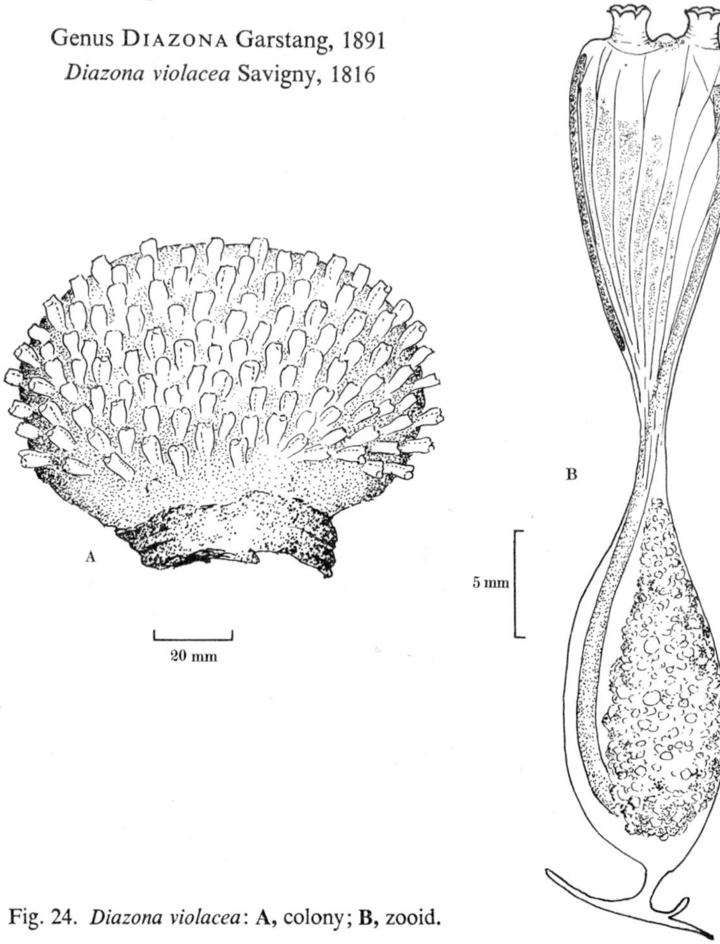

Fig. 24. *Diazona violacea*: **A,** colony; **B,** zoid.

Diazona violacea Savigny, 1816
Diazona hebridica Alder and Hancock, 1912

Colony up to 40 cm in diameter, rounded, squat or almost globular, with small projections on upper surface marking the positions of the zooids; green and translucent, soft. Zooids large, up to 5 cm long; thorax and abdomen joined by a narrow waist; siphons 6-lobed; many rows of straight stigmata; row of pointed dorsal languets; no branchial papillae; oesophagus long; stomach smooth-walled, near lower end of abdomen; rectum extends to near base of atrial siphon; ovary and testis beside intestinal loop. Oviparous.

On rocks, etc., in water of 30–200 m. Occurs on west and south coasts of British Isles.

Family PEROPHORIDAE Giard, 1872

Genus Perophora Wiegmann, 1835

Perophora listeri Forbes, 1848

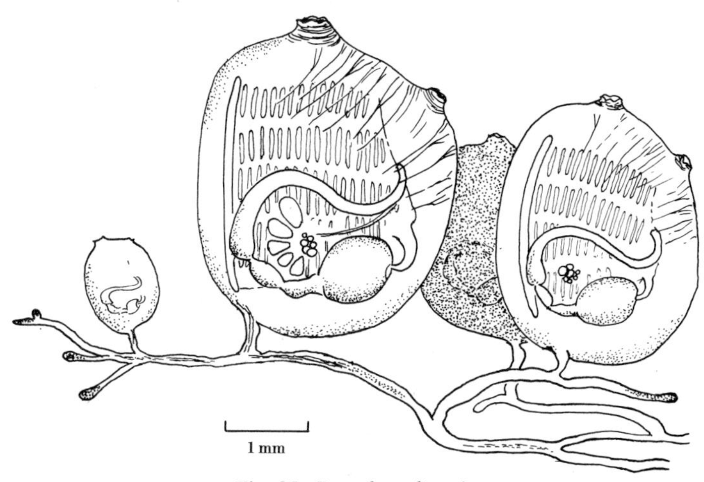

Fig. 25. *Perophora listeri.*

Perophora listeri Forbes, 1848

Colonies consist of rather widely separated zooids joined only by basal stolons which creep over the substratum. Zooids up to 4 mm high, about the same width and somewhat laterally flattened; transparent and delicate; both siphons short; 4 rows of stigmata; gut forming a loop on left side of branchial sac; gonad in intestinal loop, consisting of a small central ovary and a group of radiating testis follicles. Larvae in atrial cavity; 3 papillae in a vertical row; trunk about 0·3 mm long.

On stones, algae, etc. from the lower shore to about 30 m. Occurs on south and west coasts of the British Isles and elsewhere southwards into the Mediterranean.

Family CORELLIDAE

Genus Corella Alder and Hancock, 1870

Corella parallelogramma (Müller, 1776)

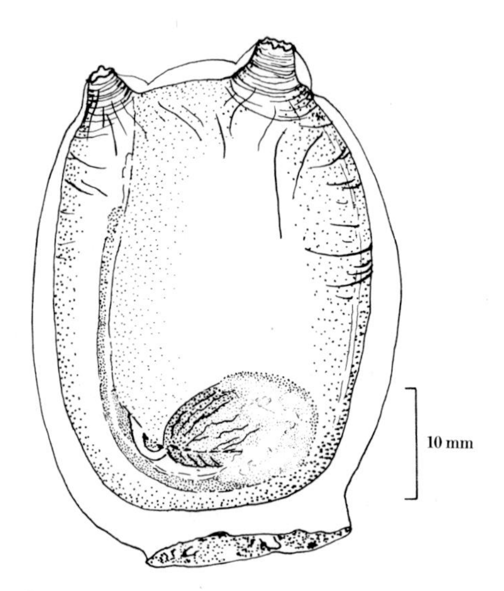

Fig. 26. *Corella parallelogramma.*

Ascidia parallelogramma Müller, 1776

Body somewhat rectangular in outline and laterally flattened; up to about 5 cm long; test smooth and almost transparent; body with yellow or red markings; stigmata spiral but divided into curved slits; no branchial papillae; gut on right side of branchial sac; stomach with longitudinal folds; intestine posterior to stomach; gonads in intestinal loop and over stomach and intestine. Oviparous.

On stones, shells, algae, etc., from shallow water down to about 200 m. Around all British coasts and elsewhere from the west coast of Norway southwards into the Mediterranean.

Kott (1951) described a new species *Corella halli* from the English Channel. Having examined the type specimen and compared it with specimens of *C. parallelogramma* of similar size, I can find no important differences and conclude that *C. halli* is a synonym of *C. parallelogramma.*

Family ASCIDIIDAE

1. Internal longitudinal branchial bars without papillae . *Ascidiella* (p. 44)

Internal longitudinal branchial bars with papillae **2**

2. Dorsal tubercle large, with one opening, posterior part of branchial sac not folded forward . . . . *Ascidia* (p. 46)

Dorsal tubercle small with several accessory openings in series behind tubercle; branchial sac folded forward . *Phallusia mammillata* (p. 52)

Genus ASCIDIELLA Roule, 1883

Body often over 6 cm long; oral tentacles fewer than internal longitudinal branchial bars of either side . . . . *A. aspersa* (p. 44)

Body usually under 4 cm long; oral tentacles more numerous than internal longitudinal branchial bars of either side . . . . *A. scabra* (p. 45)

Ascidiella aspersa (Müller, 1776)

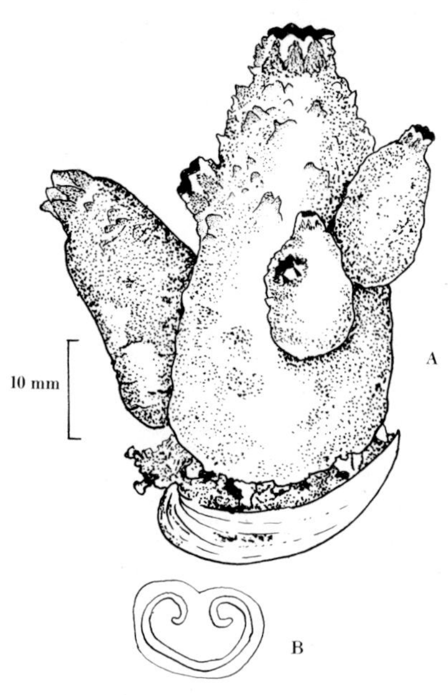

Fig. 27. *Ascidiella aspersa*: **A,** group of small individuals attached to a larger one; **B,** opening of dorsal tubercle.

Ascidia aspersa Müller, 1776
Ascidia pustulosa and *A. affinis* Alder and Hancock, 1912

Body rectangular to ovoid, somewhat narrower anteriorly; up to 10 cm long; surface of test somewhat rough; dirty grey or brown; oral siphon terminal, atrial siphon about $\frac{1}{3}$ of body length from it, on dorsal side; oral tentacles up to 40, rather widely spaced; dorsal tubercle with C-shaped slit having both horns rolled inwards; internal longitudinal branchial bars up to 80 on each side and more numerous than tentacles in any one specimen; renal vesicles large.

On stones, shells, algae, piers, etc., from low water level down to about 80 m. Generally distributed around British coasts and most common in warm sheltered areas. Also from the west coast of Norway to the Mediterranean.

Ascidiella scabra (Müller, 1776)

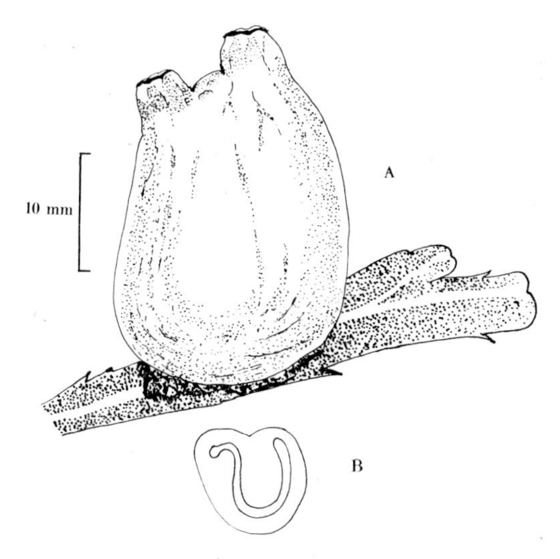

Fig. 28. *Ascidiella scabra*: **A,** individual attached to *Fucus*; **B,** opening of dorsal tubercle.

Ascidia scabra Müller, 1776
Ascidia sordida, *A. morei*, *A. normani*, *A. affinis*, *A. elliptica* and *A. pellucida* Alder and Hancock, 1912

The species differs from *A. aspersa* in: the smaller size (usually less than 4 cm long), anterior position of the atrial siphon, oral tentacles relatively closely spaced and more numerous than internal longitudinal branchial bars of either side and the red marks which are frequently present on the body.

On algae, shells, etc., from the lower shore down to 300 m. Generally distributed around Britain, extending across the North Sea to the west coast of Norway and southward into the Mediterranean.

Genus ASCIDIA Linnaeus, 1767

1. Ganglion a considerable distance behind dorsal tubercle **2**

Ganglion close to dorsal tubercle **3**

2. Test thick and smooth; body often pink; no papillae on right face of dorsal lamina . . . . *A. mentula* (p. 47)

Test thin and often rough with adhering sand, shell, etc.; body often green; papillae on right face of dorsal lamina . . . . . *A. conchilega* (p. 48)

3. Anus anterior to anterior limit of intestinal loop . . *A. virginea* (p. 49)

Anus posterior to anterior limit of intestinal loop **4**

4. No intermediate branchial papillae . . . . *A. obliqua* (p. 50)

Intermediate branchial papillae . . . . *A. prunum* (p. 51)

Ascidia mentula Müller, 1776

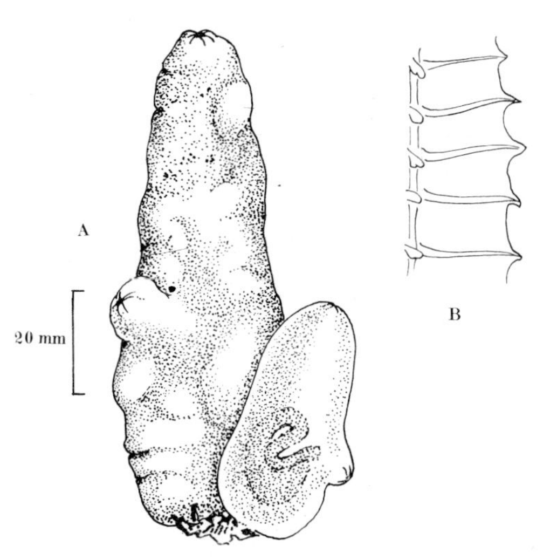

Fig. 29. *Ascidia mentula*: **A**, two individuals; **B**, part of right side of dorsal lamina.

Ascidia mentula Müller, 1776
Ascidia robusta, *A. rubicunda*, *A. rubrotincta*, *A. crassa*, *A. mollis*, *A. plana*, *A. alderi* and *A. rudis* Alder and Hancock, 1912

Body to 10 cm or more in length, surface usually smooth, with low rounded swellings; grey to pink; test thick, cartilaginous and translucent; siphons inconspicuous, the oral terminal and the atrial $\frac{1}{2}$–$\frac{2}{3}$ of body length from it; dorsal tubercle with horse-shoe shaped opening; ganglion some distance behind dorsal tubercle; branchial papillae on longitudinal bars at intersections with transverse bars and also intermediate papillae between transverse bars; right face of dorsal lamina lacks papillae; rectum short, with anus behind anterior limit of intestinal loop; lobed ovary in intestinal loop and diffuse testis on surface of intestine; renal vesicles of moderate size, on surface of stomach and intestine.

On rock, stones, shells, etc., from the lower shore to depths of 200 m. Generally distributed around British Isles; elsewhere from western Norway southwards to the Mediterranean and Black Sea.

Ascidia conchilega Müller, 1776

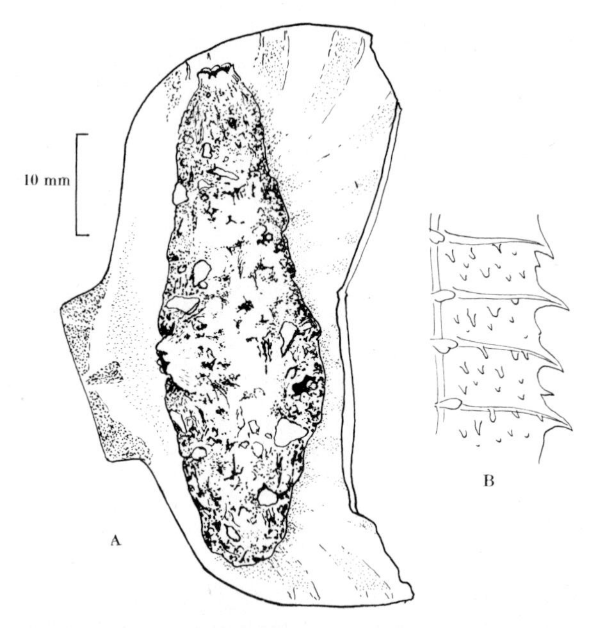

Fig. 30. *Ascidia conchilega*: **A,** individual attached to a piece of scallop shell; **B,** part of right side of dorsal lamina.

Ascidia conchilega Müller, 1776
Ascidia producta, *A. inornata*, *A. depressa*, *A. elongata*, *A. amoena* and *A. plebeia* Alder and Hancock, 1912

This species is distinguished from *A. mentula* by: its smaller maximum size (usually under 6 cm long), attachment by a large part of the left side; test thin; often rough and with adhering shell, etc., colour often greenish, right face of dorsal lamina with small irregular papillae, renal vesicles large (up to 2 mm in diameter).

On stones, shell, etc., from the lower shore to depths of at least 1000 m. Distribution similar to that of *A. mentula*.

Ascidia virginea Müller, 1776

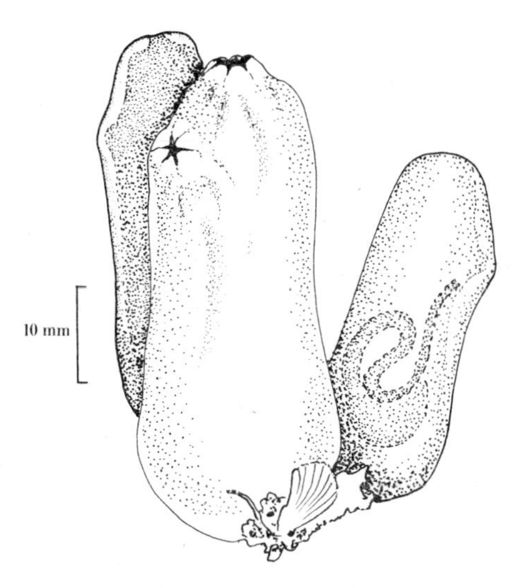

Fig. 31. *Ascidia virginea.*

Ascidia virginea Müller, 1776
Ascidia patoni Herdman, 1881
Ascidia venosa Alder and Hancock, 1905

Body somewhat rectangular in outline, up to 8 cm long; colour translucent grey to pink; test smooth; oral siphon terminal and atrial siphon a short distance from it; oral tentacles numerous (may exceed 100); rectum long, with anus anterior to anterior limit of intestinal loop.

On stones, shells, etc., from shallow water down to about 400 m. Sparsely distributed along west and south coasts of British Isles and northwards to Spitzbergen and southwards into the Mediterranean.

Ascidia obliqua Alder, 1863

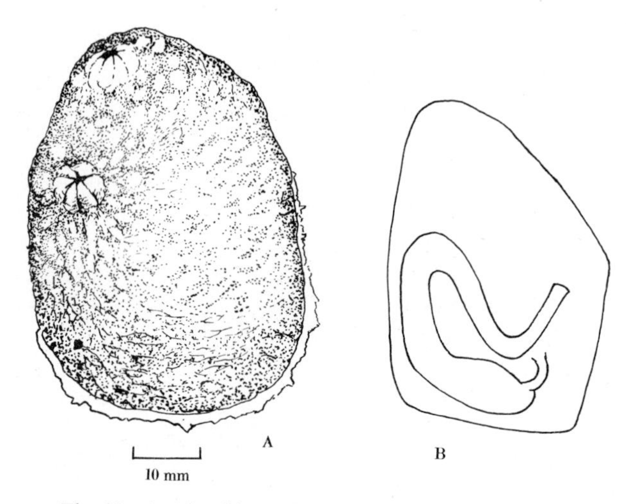

Fig. 32. *Ascidia obliqua*: **A,** individual; **B,** shape of gut.

Ascidia obliqua Alder, 1863
Ascidia falcigera and *A. mollis* Herdman, 1891
Ascidia lurida and *A. gelatinosa* Hartmeyer, 1903

Body wide and ovoid, up to 8 cm long, attached by much of the left side and somewhat flattened; oral siphon near the anterior end and atrial siphon $\frac{1}{3}$–$\frac{1}{2}$ of body length from it; test soft and rather rough; oral tentacles up to 70 in number; opening of dorsal tubercle U-shaped; branchial papillae large; no intermediate papillae; anterior part of intestinal loop not bent dorsally.

On rock, stones, shells, etc., from shallow water down to at least 1000 m. In British waters apparently only around the Shetland Islands. This is a northern species, extending along the Norwegian coast into the Arctic seas of Europe, Asia and Greenland.

Ascidia prunum Müller, 1776

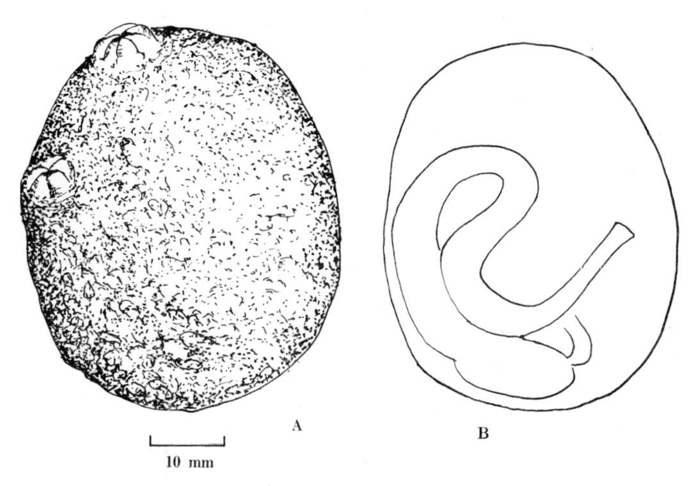

Fig. 33. *Ascidia prunum*: **A,** individual; **B,** shape of gut.

Ascidia prunum Müller, 1776 (but see Hartmeyer, 1924 and Berrill, 1950)

Body ovoid and somewhat flattened, up to 6 cm or more in length; attached by left side; test fairly smooth; oral siphon nearly terminal; atrial siphon $\frac{1}{3}$–$\frac{1}{2}$ of body length from it; oral tentacles 40–50; opening of dorsal tubercle U-shaped; intermediate branchial papillae present; anterior part of intestinal loop bent dorsally.

On stones, rock and shell from shallow water down to about 400 m. Distribution similar to that of *A. obliqua* and, like that species, recorded in British waters only from the Shetland Islands.

Genus PHALLUSIA Savigny, 1816

Phallusia mammillata (Cuvier, 1815)

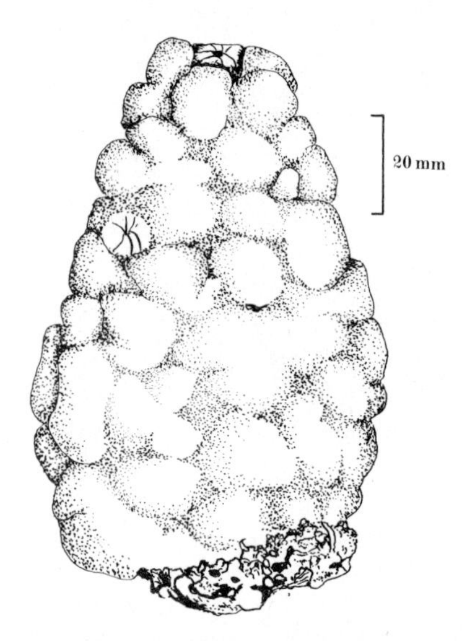

Fig. 34. *Phallusia mammillata*.

Ascidia mammillata Cuvier, 1815

Body ovoid, usually wider at the base, up to 14 cm long; surface with numerous large rounded smooth swellings; test very thick and cartilaginous; oral siphon terminal, atrial siphon $\frac{1}{3}$–$\frac{1}{2}$ of body length from it; oral tentacles up to 100; dorsal tubercle small but numerous small secondary openings are situated between the tubercle and the ganglion; intermediate branchial papillae present; posterior part of branchial sac is folded forwards; gut a compact mass with anus about level of anterior limit of intestinal loop.

On stones, etc., from the lower shore down to about 180 m.

Known with certainty in British waters only from the south coast; more northerly records require confirmation. Further distribution mainly in the Mediterranean.

Family STYELIDAE

1. Solitary **2**

Colonial **6**

2. Each gonad consists of central tubular ovary and male follicles beside but separate from ovary **3**

Each gonad consists of ovary and male follicles bound together into a compact organ. **4**

3. Branchial sac with folds (sometimes reduced); gonads not U-shaped *Styela* (p. 55)

Branchial sac without folds; one U-shaped gonad on each side *Pelonaia* (p. 54)

4. Gonad on right side only . . . . *Dendrodoa* (p. 64)

Gonads on both sides **5**

5. Gonads elongated, in a row on each side . . . *Cnemidocarpa* (p. 58)

Gonads short, many and usually irregularly arranged . *Polycarpa* (p. 59)

6. Zooids not arranged in systems **7**

Zooids arranged in systems. **9**

7. Stigmata longitudinal (i.e. parallel to endostyle) **8**

Stigmata transverse (i.e. at right angles to endostyle) *Protostyela* (p. 72)

8. Zooids more or less separate, united by basal stolons; branchial sac with distinct folds . . . . *Stolonica* (p. 68)

Zooids closely packed; branchial sac without true folds . *Distomus* (p. 66)

9. Zooids in regular star-shaped systems . . . . . . *Botryllus* (p. 70)

Zooids in elongated systems. . . . . *Botrylloides* (p. 71)

Genus PELONAIA Forbes and Goodsir, 1841

Pelonaia corrugata Forbes and Goodsir, 1841

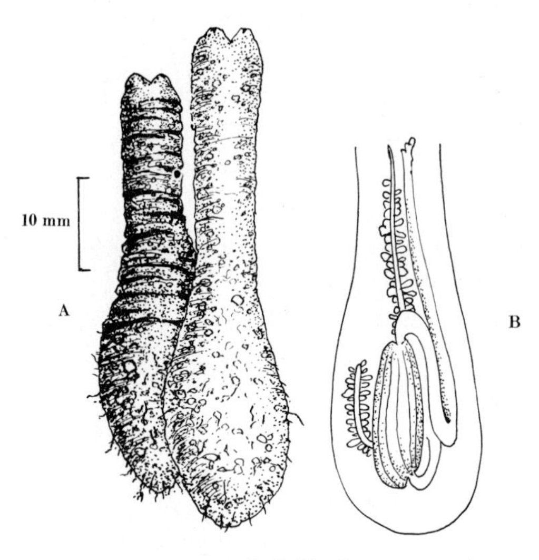

Fig. 35. *Pelonaia corrugata*: **A,** individuals; **B,** gut and one gonad.

Pelonaia corrugata Forbes and Goodsir, 1841
Pelonaia glabra Alder and Hancock, 1907

Body up to 14 cm long, elongated and club-shaped, the lower swollen end partially embedded in the substratum; surface coated with sand, mud, etc.; siphons small and close together at upper end; branchial sac without folds; stomach more or less longitudinal in body, intestine bent back, then forward, rectum long; one U-shaped gonad on each side, consisting of tubular ovary and a series of small testis follicles along each side of ovary. Oviparous, but development direct without a tadpole stage.

On bottoms of sand and mud. In British waters confined to northern areas, on the west as far south as the Firth of Clyde and on the east as far south as the north-east coast of England. Its range extends northwards into Arctic seas.

1. Body elongated with distinct narrow stalk attached at its lower end to substratum . . . . *S. clava* (p. 57)

Body not elongated; no stalk **2**

2. Two gonads on each side . . . . *S. partita* (p. 56)

One gonad on each side . . . . *S. coriacea* (p. 55)

Styela coriacea (Alder and Hancock, 1848)

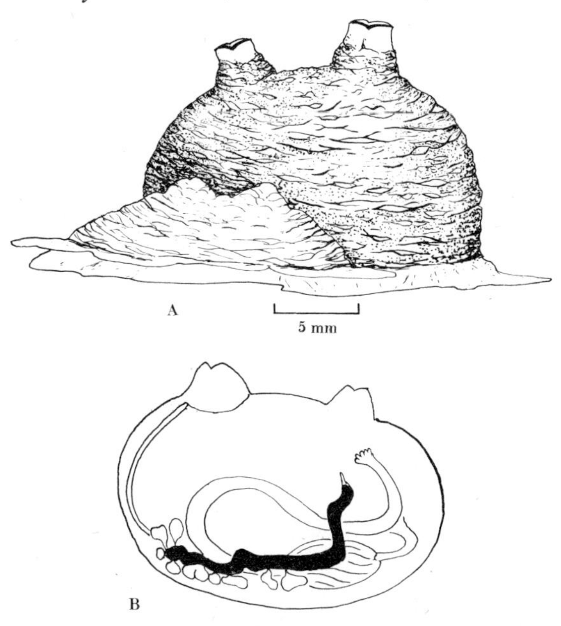

Fig. 36. *Styela coriacea*: **A,** individual; **B,** shape of gut and one gonad.

Cynthia coriacea Alder and Hancock, 1848
Styela granulata and *S. northumbrica* Alder and Hancock, 1907
Styela loveni Ärnbäck-Christie-Linde, 1922

Body low and dome-shaped to nearly cylindrical, up to 2 cm in diameter; surface minutely wrinkled; siphons rather far apart on upper side of body; 4 quite large branchial folds on each side; stomach horizontal, spindle-shaped and with longitudinal folds, intestine and rectum forming an S-shaped curve; one gonad on each side, with sinuous or L-shaped ovary and groups of testicular follicles mainly along the lower sides of ovaries. Oviparous and perhaps sometimes larviparous.

On shells, stones, the tests of other ascidians, etc., from shallow water down to about 600 m. Generally distributed around British coasts but apparently scarce or absent on much of the east coast. The species occurs northwards along the European and north-east American coasts into the Arctic.

Styela partita (Stimpson, 1852)

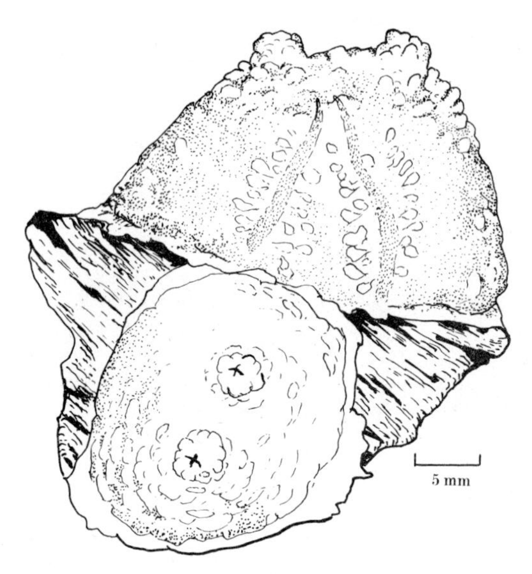

Fig. 37. *Styela partita*: the upper individual with test partly removed to show gonads of one side.

Cynthia partita Stimpson, 1852

Body seldom over 2 cm long, ovoid or somewhat conical with expanded base; siphons fairly close together on upper side; surface somewhat rough with rounded projections, often more conspicuous around the siphons; pale dirty yellow to brown; four branchial folds on each side; shape of stomach intestine and rectum much as in *S. coriacea* but with the loops less dorso-ventrally flattened; two long ovaries on each side, the ovaries being surrounded by small testis follicles. Oviparous.

Attached to rocks, stones, etc., in shallow water down to about 30 m. In British waters known only from parts of the south coast. Wider distribution extends southwards into the Mediterranean, on the west coast of Africa and on the east coast of the U.S.A.

Styela clava Herdman, 1882

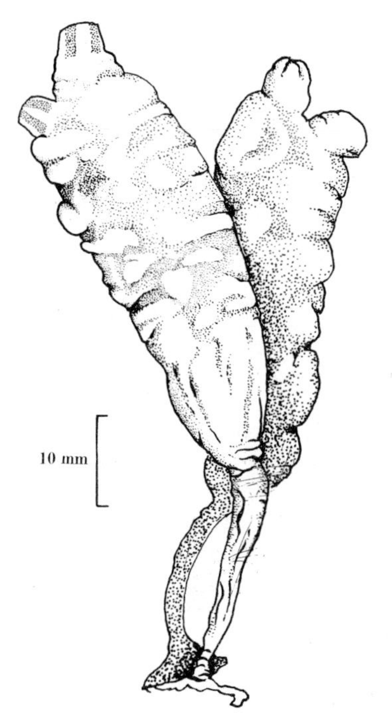

Fig. 38. *Styela clava.*

Styela clava Herdman, 1882
Styela mammiculata Carlisle, 1954

Body long and narrow, tapering back to narrow distinct stalk which may occupy $\frac{1}{3}$ of the total length of up to 12 cm; siphons close together at anterior end; surface with mammilations, folds and swellings; 4 branchial folds on each side; stomach longitudinal, with many folds, intestinal loop narrow; gonads 2–4 on left side, 5–8 on right side, each with long narrow ovary surrounded by male follicles.

On stones, harbour installations, etc., in shallow water. A species accidentally introduced, possibly from Korean waters, and at present known in British waters only from certain parts of the south coast where, however, it may be abundant. Native to Japan, the Sea of Okhotsk, Korea and Siberia.

Genus CNEMIDOCARPA Huntsman, 1912

Cnemidocarpa mollis (Stimpson, 1852)

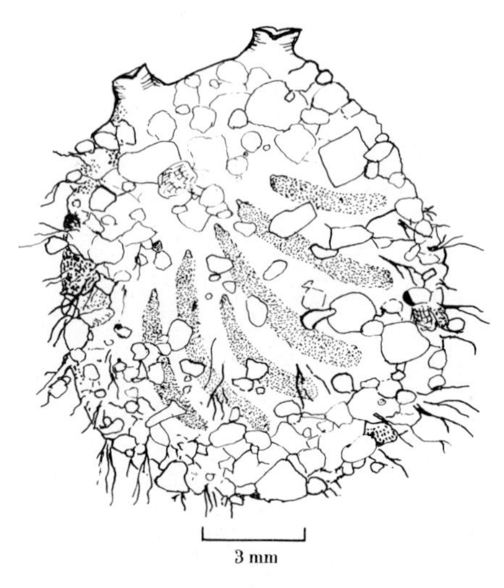

Fig. 39. *Cnemidocarpa mollis*, with gonads of one side visible through the transparent test and body wall.

Glandula mollis Stimpson, 1852
Styela vestita Alder and Hancock, 1907

Body spherical to ovoid, up to 1·5 cm in diameter; siphons fairly close together on upper side; test fibrils on lower part; surface coated with sand, broken shell, etc.; branchial folds rudimentary; stomach ovoid with longitudinal folds; intestine and rectum forming a rather simple S-bend; gonads cylindrical to club-shaped, each consisting of a central ovary and closely applied testis follicles bound together as a compact organ; 3–6 gonads on the left side and 4–8 on the right.

Loosely attached by the test fibrils or partly embedded in sand or mud, from shallow water down to at least 100 m. Known from eastern and northern Scottish waters, including the Orkney and Shetland Islands, and elsewhere from the north-east coast of the U.S.A.

Genus POLYCARPA Heller, 1877

1. Gonads in a single row on each side of endostyle *Polycarpa gracilis* (p. 61)

Gonads scattered irregularly over body wall **2**

2. Surface smooth; no test fibrils; colour usually red *Polycarpa rustica* (p. 62)

Surface rough, wrinkled or with test fibrils; body not red (or only around siphons) **3**

3. Test with fibrillar processes and heavily coated with sand, mud, etc. *Polycarpa fibrosa* (p. 63)

Test without fibrillar processes and with little or no adhering sand, etc. *Polycarpa pomaria* (p. 60)

Polycarpa pomaria (Savigny, 1816)

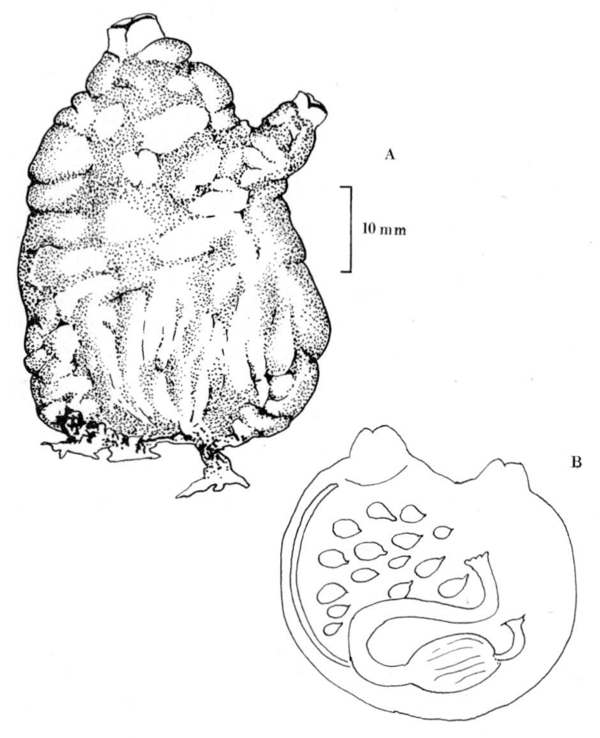

Fig. 40. *Polycarpa pomaria*: **A,** individual; **B,** gut and gonads of left side.

Cynthia pomaria Savigny, 1816
Styela tuberosa, *S. informis*, *S. quadrangularis*, *S. opalina* and *S. sulcatula* Alder and Hancock, 1907

Body ovoid, spherical or somewhat conical; up to 7 cm long; siphons, especially atrial siphon, may be quite long or relatively inconspicuous; surface wrinkled, mammilated or with large irregular swellings, usually dirty brown in colour; 4 low branchial folds on each side; stomach ovoid, with longitudinal folds, and sometimes not clearly marked off from intestine; intestine and rectum forming S-shaped bend; gonads flask-shaped to ovoid, numerous, on each side of the body.

On stones, shells, etc., from shallow water down to about 450 m. Generally distributed and often common around British coasts. Wider distribution from Spitzbergen southwards along European coasts and the North Sea, into the Mediterranean.

Polycarpa gracilis Heller, 1877

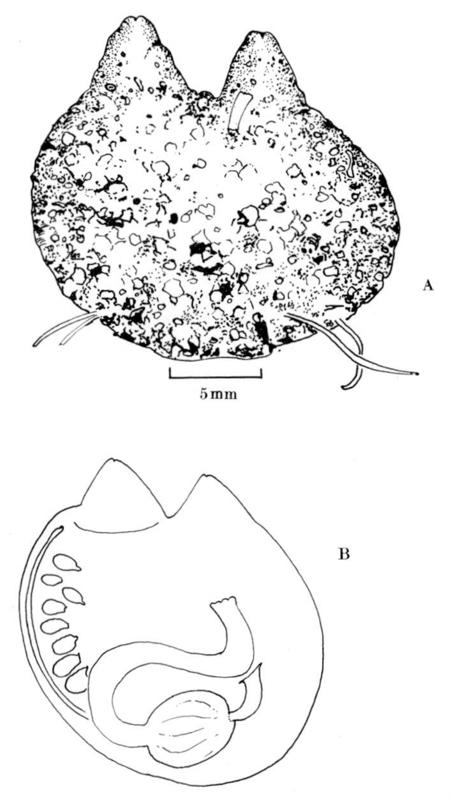

Fig. 41. *Polycarpa gracilis*: **A,** individual; **B,** gut and gonads of left side.

Polycarpa gracilis Heller, 1877
Styela humilis, *S. obscura* and *S. depressa* Alder and Hancock, 1907

Body ovoid, squat to upright, up to 2 cm long; test with fine fibrils and coated with sand, mud or shell; siphons fairly prominent; 4 branchial folds on each side; stomach ovoid, with longitudinal folds; intestine and rectum forming a simple S-shaped loop; gonads ovoid to flask-shaped, arranged in a row on each side of the endostyle.

On stones, etc., in shallow water. In the British Isles known from waters of the English Channel and parts of the south coast and possibly from the west coast of Scotland. Further distribution extends through the Mediterranean to the Adriatic.

Polycarpa rustica (Linnaeus, 1767)

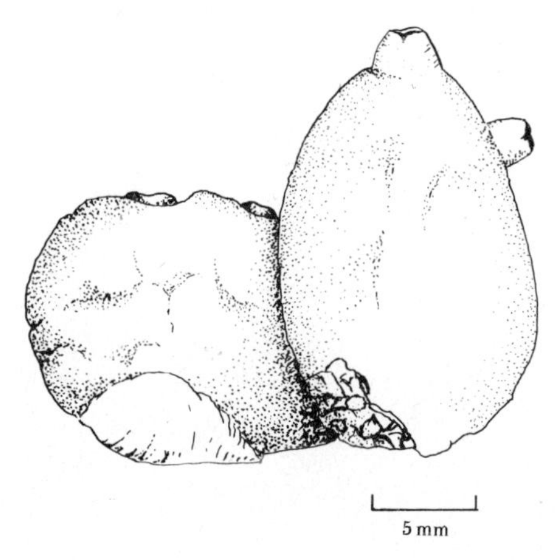

Fig. 42. *Polycarpa rustica*, one expanded and one contracted individual.

Ascidia rustica Linnaeus, 1767

Body pink to red, ovoid and upright to almost spherical, up to about 2 cm high; oral siphon terminal and atrial siphon a short distance from it on dorsal side; test fairly smooth; 4 branchial folds on each side; gut and gonads much as in *P. pomaria*.

On stones, etc., from the lower shore to shallow water. Known from the south and west coasts of the British Isles, as far north as south-west Scotland, and elsewhere apparently only from the English Channel.

Polycarpa fibrosa (Stimpson, 1852)

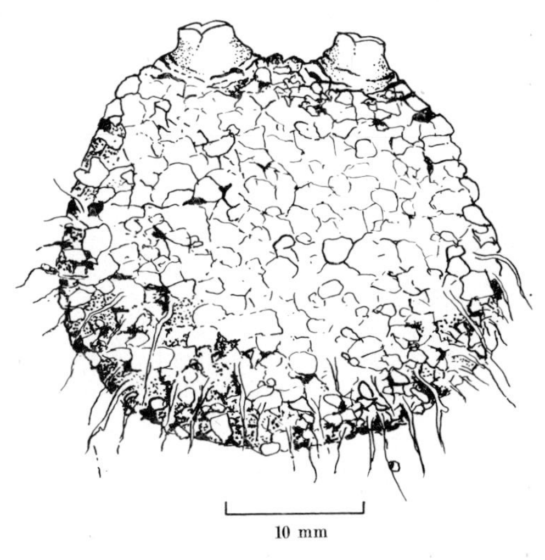

Fig. 43. *Polycarpa fibrosa.*

Glandula fibrosa Stimpson, 1852
Styela comata and *S. fibrillata* Alder and Hancock, 1907
Polycarpa libera Ärnbäck-Christie-Linde, 1922

Body spherical to ovoid, up to about 3 cm in diameter, test with fibrils and a coating of sand, mud, or shell, except for a bare area on or round the siphons; 4 branchial folds on each side; gut and gonads much as in *P. pomaria.*

Loosely attached to sand or mud, from low water level to depths of over 1000 m. Generally distributed, but not common, round all British coasts, also northwards along European coasts into the Arctic, and on the north-east coasts of America.

Genus DENDRODOA MacLeay, 1825

Dendrodoa grossularia (van Beneden, 1846)

Ascidia grossularia van Beneden, 1846
Styelopsis grossularia and *S. sphaerica* Alder and Hancock, 1907

Body varies from depressed and dome-shaped to upright and cylindrical; up to 2 cm long; red or brown; branchial sac without true folds but groups of internal longitudinal bars, or with one low dorsal fold on the right side; stomach horizontal with longitudinal folds; intestine and rectum forming a simple, often horizontally flattened S-shaped bend; one gonad on right side and none on left; gonad a straight cylindrical body with central ovary and testis follicles on lower side; larviparous; eggs large and red; larvae, in posterior part of atrial cavity, trunk nearly spherical, with 3 small papillae, ring of many anterior ampullae and single black sensory organ (otolith).

On stones, rock, shells, algae, the test of other ascidians, etc., from the lower shore down to depths of at least 600 m. Generally distributed, and locally common, around the British coasts; along European coasts to Arctic waters and on the north east coast of America.

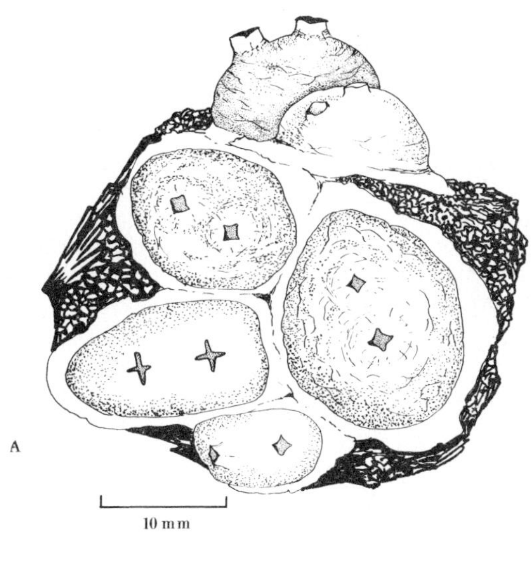

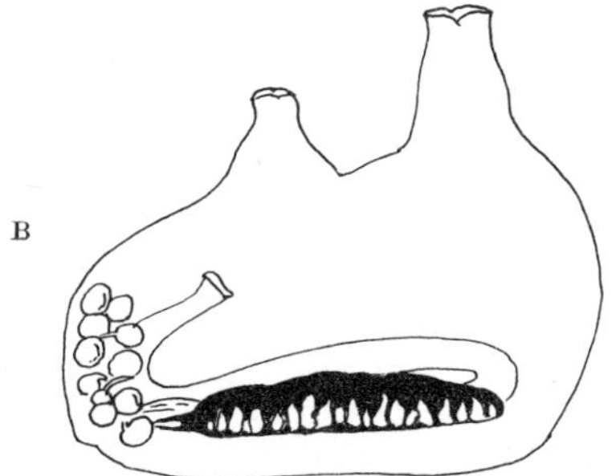

Fig. 44. *Dendrodoa grossularia*: **A,** group of individuals; **B,** gut, gonad and group of developing eggs.

Genus DISTOMUS Gaertner, 1774

Distomus variolosus Gaertner, 1774

Distomus variolosus Gaertner, in Pallas, 1774

Zooids closely grouped but united only by their bases or sides; cylindrical to ovoid; red to brown; test somewhat rough; siphons both at upper end; body up to about 1 cm long; branchial folds rudimentary, sometimes one dorsal fold on each side, sometimes represented only by groups of internal longitudinal bars; stomach with folds; ovaries on right side only; each with one large egg; testes ovoid or elongated, on left side only. Larviparous; larvae with 3 simple papillae in triangular arrangement, a ring of anterior ampullae and a black sensory organ (photolith) which is a combined ocellus and otolith; trunk about 0·7 mm long.

On rock, stones, algae, etc., from low water level to shallow water. Recorded from the south coast of England; records from farther north, along the west coast, require confirmation because this species may have been confused with *Dendrodoa grossularia* which it may resemble externally.

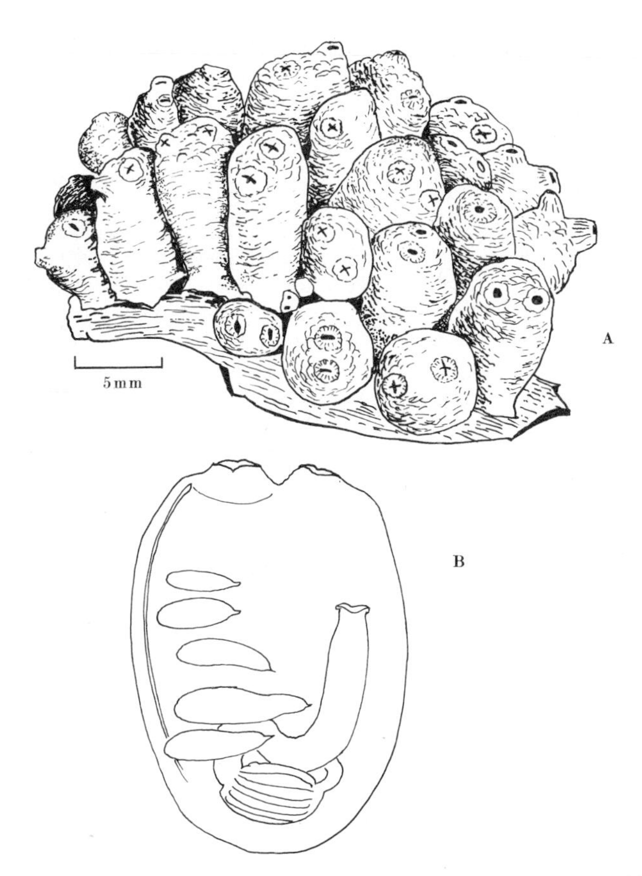

Fig. 45. *Distomus variolosus*: **A,** colony; **B,** gut and left gonads of a zooid.

Genus Stolonica Lacaze-Duthiers and Delage, 1892

Stolonica socialis Hartmeyer, 1903

Stolonica socialis Hartmeyer, 1903
Thylacium aggregatum Alder and Hancock, 1907

Zooids separate and joined by creeping basal stolons; up to 2 cm in height, yellow, orange or brown, ovoid or somewhat rectangular in outline, siphons small and close together on upper end of zooid, test fairly smooth with little adhering sand; stolons coated with sand, etc.; 2 or 3 branchial folds on right side and 3 (or 2) on left; stomach more or less vertical, with folds, intestine with sharp bend at stomach, then with rectum almost straight; gonads in 3 rows, one on each side of the endostyle and one parallel to the intestine; hermaphrodite gonads confined to posterior part of left endostylar row, all remaining gonads male and each consisting of a rosette of small follicles. Larviparous; larva somewhat like that of *Distomus variolusus* but trunk about 1 mm long.

On rock and stones, from just below low tide level to about 35 m. In Britain only on south and south-west coasts and further distribution appears to be limited to the English Channel.

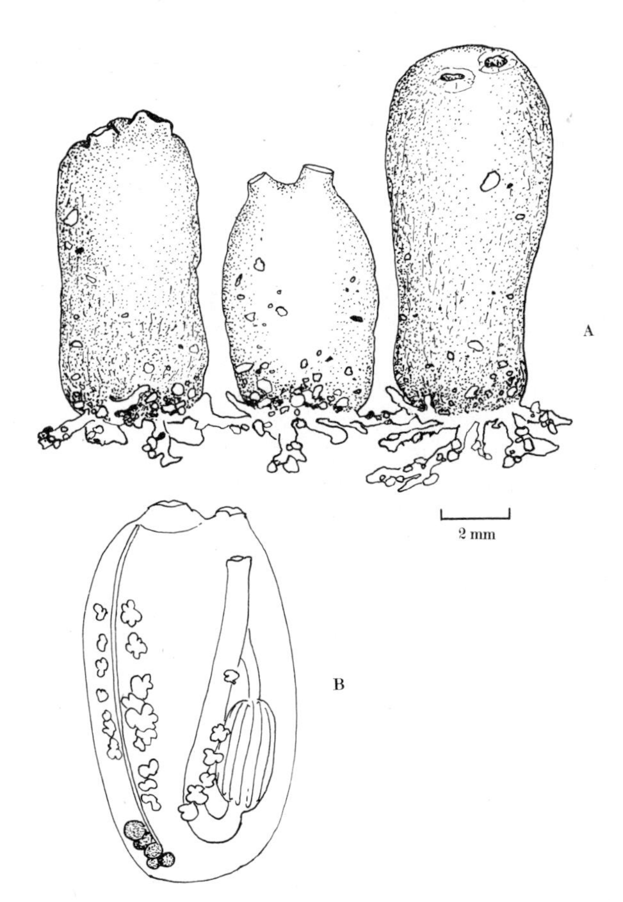

Fig. 46. *Stolonica socialis*: **A,** colony; **B,** zooid from left ventral side, to show gut and gonads.

Genus BOTRYLLUS Gaertner 1774

Botryllus schlosseri (Pallas, 1766)

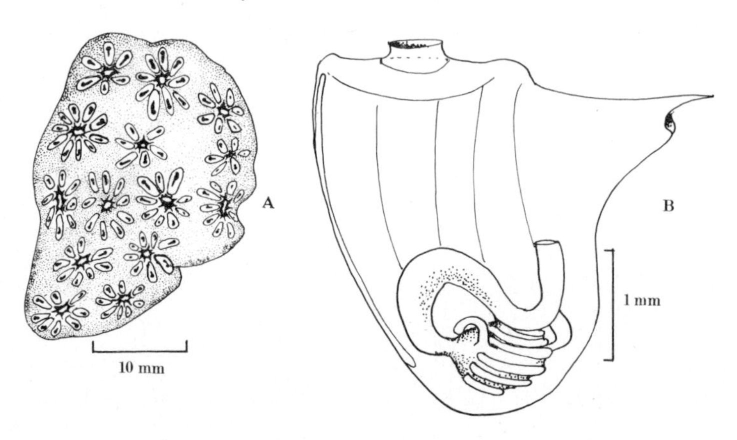

Fig. 47. *Botryllus schlosseri*: **A,** colony; **B,** zooid.

Alcyonium schlosseri Pallas, 1766
Botryllus polycyclus, *B. gemmeus* and *B. minutus* Savigny, 1816
Botryllus smaragdus and *B. bivittatus* Milne-Edwards, 1842
Botryllus badium, *B. calyculatus* and *B. miniatus* Alder and Hancock, 1912

Colony flat or fleshy; zooids arranged in conspicuous star-shaped systems, each with a central common cloacal opening; colours vary greatly and include green, violet, brown and yellow, the zooids contrasting with the test. Zooids have smooth-rimmed oral siphon and large atrial siphon with projecting upper border; no branchial folds; 3 longitudinal branchial bars on each side; stomach barrel-shaped, with hooked pyloric caecum; intestine and rectum forming S-shaped bend; one gonad on each side, consisting of a group of rounded or pear-shaped male follicles, and an ovary, with a few large eggs, dorsal to testis. Larvae in atrial cavity; trunk about 0·5 mm long with 3 simple papillae in triangular arrangement, ring of anterior ampullae, black photolith.

On algae, rock, stones, the test of other ascidians, etc., from the lower shore to shallow water but occasionally to several hundred metres. Generally distributed, and sometimes common in British waters; from northern Norway along European coasts and through the Mediterranean.

Genus BOTRYLLOIDES Milne-Edwards, 1841

Botrylloides leachi (Savigny, 1816)

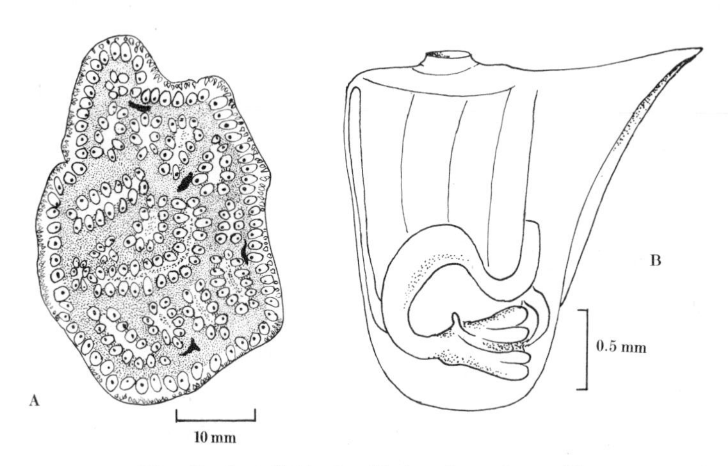

Fig. 48. *Botrylloides leachi*: **A,** colony; **B,** zooid.

Botryllus leachi Savigny, 1816
Botrylloides vinosa, *B. rubrum*, *B. albicans*, *B. radiata*, *B. ramulosa*, *B. sparsa* and *B. pusilla* Alder and Hancock, 1912

Colony flat, brown to orange-red; zooids in long narrow systems, in contrast to *Botryllus schlosseri*. Zooids differ from those of *B. schlosseri* in: stomach wider at oesophageal end than intestinal end; pyloric caecum small; ovary posterior to testis. Larvae in atrial cavity; similar to those of *Botryllus schlosseri*.

On algae, stones, the test of other ascidians, etc., from the lower shore to shallow water but occasionally in much deeper water. Generally distributed around the British Isles and further geographical range similar to that of *Botryllus schlosseri*.

Genus PROTOSTYELA Millar, 1954

Protostyela heterobranchia Millar, 1954

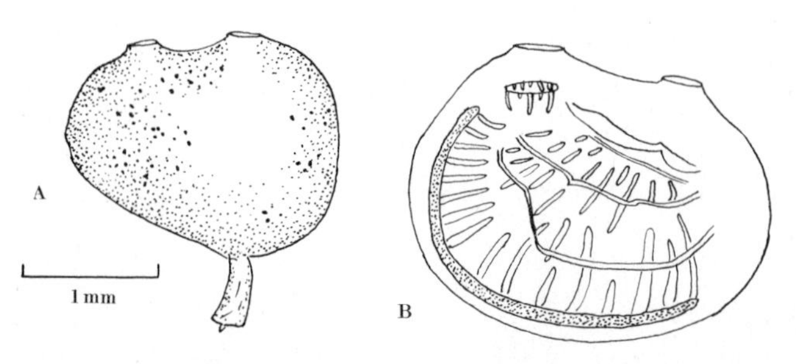

Fig. 49. *Protostyela heterobranchia*: **A,** zooid; **B,** branchial sac to show transverse stigmata and longitudinal bars.

Protostyela heterobranchia Millar, 1954

The species is known only from the type specimen; zooid 2 mm long, with separate oral and atrial siphons, no branchial folds, 3 longitudinal branchial bars on each side, stigmata transverse (at right angles to endostyle); stomach barrel-shaped, with folds, intestine and rectum in S-shaped bend; 3 gonads on right side, the anterior one hermaphrodite and the posterior ones male; one hermaphrodite gonad on left side. Larvae in atrial cavity. Larva (possibly not fully developed) with trunk about 0·22 mm long, 3 simple papillae in triangular arrangement and a single black sense organ.

The only specimen was collected from low water level in Argyll, Scotland.

Family PYURIDAE

1. Surface with large branched spines; stigmata transverse *Boltenia echinata* (p. 73)

Surface without large branched spines; stigmata longitudinal . . . **2**

2. Dorsal lamina an undivided membrane . . . . *Microcosmus* (p. 74)

Dorsal lamina represented by a series of languets . . . *Pyura* (p. 75)

Genus BOLTENIA Savigny, 1816

Boltenia echinata (Linnaeus, 1767)

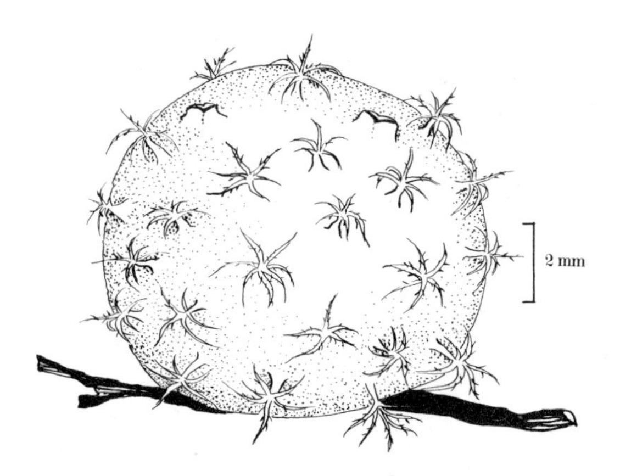

Fig. 50. *Boltenia echinata.*

Ascidia echinata Linnaeus, 1767

Body spherical, up to 2 cm in diameter; surface with large conspicuous branched spines; siphons short, on upper side of body; 6 branchial folds (occasionally 7 or 8) on each side; primary intestinal loop narrow, secondary loop widely open; one gonad on each side, the left gonad being in the primary intestinal loop; gonads long and lobed. Larvae in atrial cavity.

Attached to the test of ascidians, hydroids, worm-tubes, stones or shells, from shallow water to about 350 m.

Generally distributed around the coasts of Scotland, the northern parts of the Irish Sea and north east England. The further distribution extends northwards into the Arctic and includes north east North America.

Genus Microcosmus Heller, 1877

Microcosmus claudicans (Savigny, 1816)

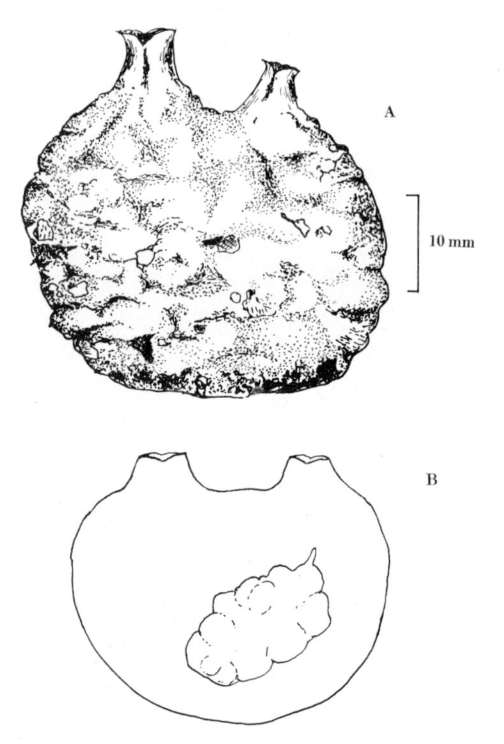

Fig. 51. *Microcosmus claudicans*: **A,** individual; **B,** gonad of left side.

Cynthia claudicans Savigny, 1816
Cynthia rosea Alder and Hancock, 1912

Body spherical to ovoid, up to 4 cm in diameter; siphons quite large; sand, shell, etc., generally attached to surface; 7 or 8 branchial folds on the right side and 8 or 9 on the left; primary intestinal loop narrow, secondary loop widely open; one gonad on each side, the left gonad covering much of the intestine; gonads large and sac-like.

On rock and stones, in shallow water and down to about 40 m. East and south coasts of England; one doubtful record for west coast of Scotland.

Genus PYURA Molina, 1782

1. Surface conspicuously tessellated; 4 branchial folds on each side *Pyura tessellata* (p. 75)

   Surface not tessellated; more than 4 branchial folds on each side . . 2

2. Six branchial folds on each side; one gonad on each side *Pyura squamulosa* (p. 76)

   Seven branchial folds on each side; 2 gonads on right side, one on left *Pyura microcosmus* (p. 77)

Pyura tessellata (Forbes, 1848)

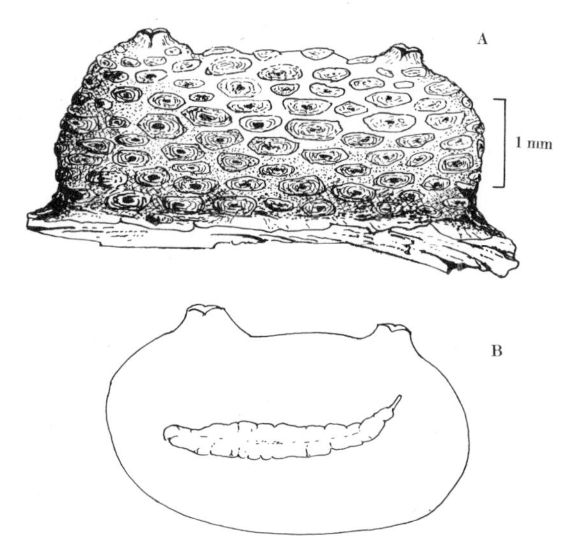

Fig. 52. *Pyura tessellata*: **A,** individual; **B,** gonad of left side.

Cynthia tessellata Forbes, 1848

Body usually dorso-ventrally flattened, up to 1 cm long, and usually brown; surface of test divided into hexagonal or rectangular plates with concentric lines, giving the body a conspicuously tessellated appearance; siphons far apart; 4 branchial folds on each side (occasionally only 3 on one side); primary intestinal loop narrow, secondary loop widely open; one long narrow gonad on each side, the left gonad being within the primary intestinal loop. Oviparous.

On stones, rock, the test of ascidians, etc., from the lower shore to about 300 m. West and south coasts of the British Isles and further distribution from west coast of Norway to the Mediterranean.

Pyura squamulosa (Alder, 1863)

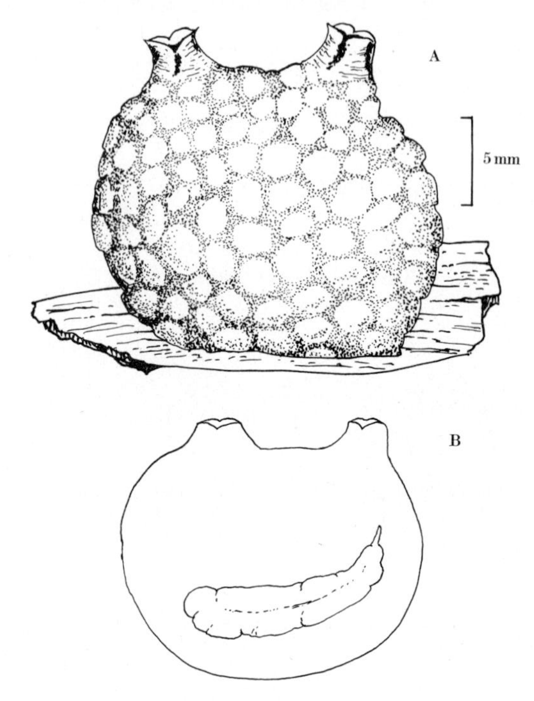

Fig. 53. *Pyura squamulosa*: **A,** individual; **B,** gonad of left side.

Cynthia squamulosa Alder, 1863
Cynthia dura Heller, 1877
Cynthia ovata Alder and Hancock, 1907

Body spherical to ovoid, up to about 2·5 cm in diameter; siphons closer together than in *P. tessellata*: surface with small rounded plates, but not giving the body a conspicuously tessellated appearance as in *P. tessellata*; 6 branchial folds on each side; intestine and gonads similar to those of *P. tessellata*. Oviparous.

On rock, stones, etc., from the lower shore to shallow water. North, west and south coasts of the British Isles. Further distribution southwards into the Mediterranean.

Pyura microcosmus (Savigny, 1816)

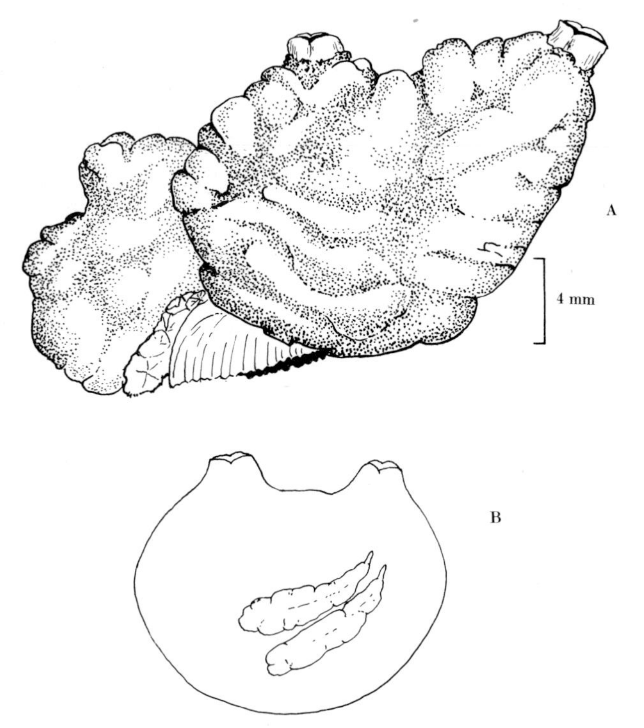

Fig. 54. *Pyura microcosmus*: **A,** two individuals; **B,** gonads of right side.

Cynthia microcosmus Savigny, 1816
Pyura savignyi Thompson, 1930

Body ovoid to elongated, with oral siphon terminal and atrial siphon some distance back along dorsal side; up to 3 cm long; usually reddish; surface with tubercles; 7 branchial folds on each side; primary intestinal loop rather widely open; one gonad on left side, within primary intestinal loop; 2 parallel gonads on right side. Oviparous.

On stones, shells, etc., in shallow water. On west and south coasts of the British Isles and extending southwards into the Mediterranean.

Family MOLGULIDAE

Branchial sac with folds . . . . *Molgula* (p. 78)

Branchial sac without folds . . . . *Eugyra* (p. 84)

Genus MOLGULA Forbes, 1848

1. Branchial sac with 6 folds on each side; oviduct directed posteriorly *Molgula manhattensis* (p. 79)

Branchial sac with 7 folds on each side (occasionally *M. complanata* has only 6) oviduct directed posteriorly (or anteriorly *M. complanata* only) . **2**

2. Branchial folds more or less reduced; oviduct directed anteriorly *Molgula complanata* (p. 83)

Branchial folds well developed; oviduct directed posteriorly . . . **3**

3. Oviduct long and narrow; test not coated with sand, shell, etc.; larviparous *Molgula citrina* (p. 82)

Oviduct of moderate length; test heavily coated with sand, shell, etc.; oviparous **4**

4. Body up to 8 cm diameter; bare area around and between siphons *Molgula oculata* (p. 80)

Body up to 3 cm diameter; no conspicuous bare area around and between siphons . . . . *Molgula occulta* (p. 81)

Molgula manhattensis (De Kay, 1843)

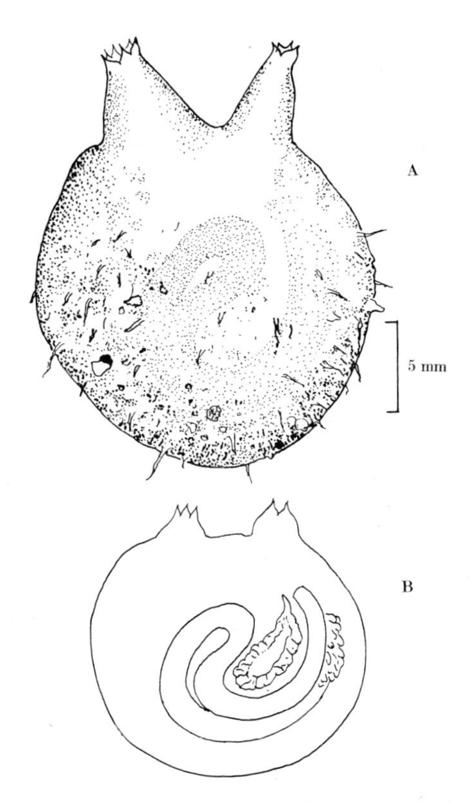

Fig. 55. *Molgula manhattensis*: **A,** individual; **B,** gut and left gonad.

Ascidia manhattensis De Kay, 1843
Molgula socialis, *M. simplex* and *M. ampulloides* Lacaze-Duthiers, 1877
Molgula siphonata and *M. inconspicua* Alder and Hancock, 1907
Molgula tubifera Ärnbäck-Christie-Linde, 1928

Body more or less spherical, up to 3 cm in diameter; grey to greenish; siphons quite large; test with fibrils and some adhering sand, etc.; 6 branchial folds; primary intestinal loop very narrow, secondary loop deeply C-shaped; one club-shaped gonad on each side, having central ovary and fringing male follicles; left gonad in secondary intestinal loop. Oviparous.

Attached to rock, stones, algae, piers, etc., or apparently loosely attached on sandy bottoms; from the lower shore to about 90 m.

Generally but unevenly distributed around British coasts. Further distribution from northern Norway to Portugal and north-east American waters.

Molgula oculata Forbes, 1848

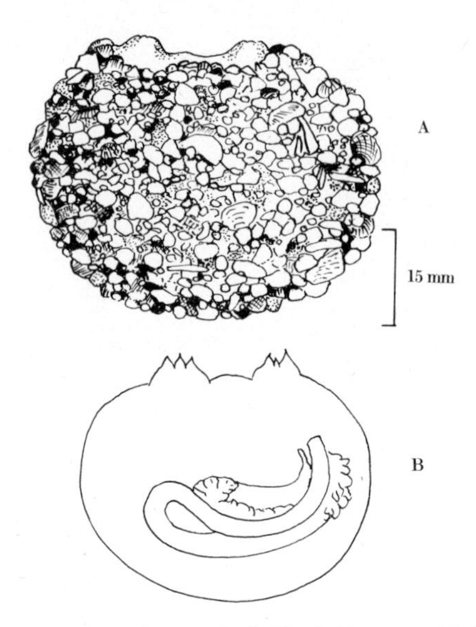

Fig. 56. *Molgula oculata*: **A,** individual; **B,** gut and left gonad.

Molgula oculata Forbes, 1848

Body ovoid to globular, up to 8 cm in diameter; heavily coated with sand, shell, etc., except for a bare area around and between the siphons, this area usually forming a depression in preserved specimens; 7 branchial folds on each side; primary intestinal loop narrow, secondary loop widely open; one gonad on each side, with central ovary and marginal male follicles; oviduct of moderate length and turned upwards. Oviparous, with formation of tadpole larva.

Unattached, on sand or gravel, from about low tide level to about 80 m. Recorded in British waters from Shetland Islands, north-east coast of England and the English Channel. The species extends southwards to the Bay of Biscay.

Molgula occulta Kupffer, 1875

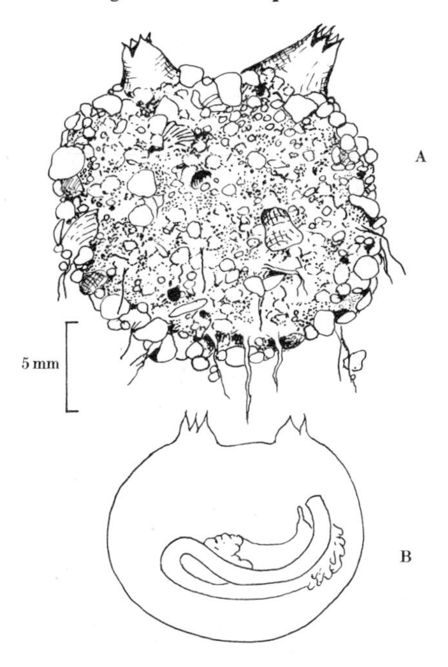

Fig. 57. *Molgula occulta*: **A,** individual; **B,** gut and left gonad.

Molgula occulta Kupffer, 1875

Body ovoid to globular, up to 3 cm in diameter, heavily coated with sand, shell or mud; no conspicuous bare area around siphons but the siphons themselves are bare; 7 branchial folds on each side; gut and gonads similar to those of *M. oculata*. Oviparous but no tadpole larva formed, development being direct.

Unattached, on sand, mud or gravel, from shallow water to about 100 m. On west and south coasts of the British Isles. Further distribution from west coast of Norway to the Mediterranean.

Molgula citrina Alder and Hancock, 1848

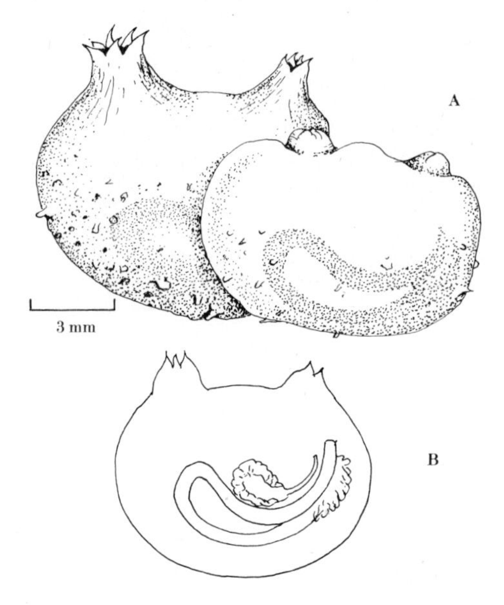

Fig. 58. *Molgula citrina*: **A,** two individuals; **B,** gut and left gonad.

Molgula citrina Alder and Hancock, 1848

Body globular or somewhat depressed, up to about 1·5 cm in diameter, grey-green; surface usually bare but may have some sand, etc., adhering; 7 branchial folds on each side; primary intestinal loop narrow, secondary loop almost semi-circular; one gonad on each side; left gonad in secondary intestinal loop, gonads ovoid, with long narrow oviduct. Larviparous.

Attached to stones, shells, algae, etc., from the lower shore to shallow water, but occasionally down to 200 m. Generally distributed around the British Isles, and extending northwards along the coast of Europe into Arctic waters, and also on the north-east coast of North America.

Molgula complanata Alder and Hancock, 1870

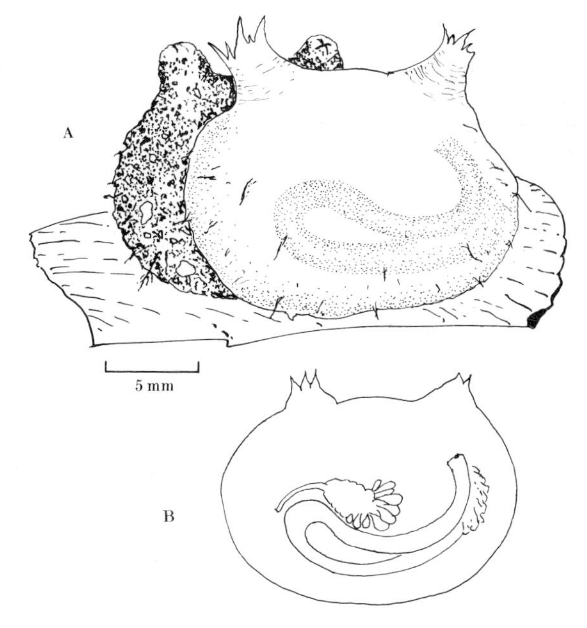

Fig. 59. *Molgula complanata*: **A,** two individuals, one with test cleaned; **B,** gut and left gonad.

Molgula complanata Alder and Hancock, 1870
Lithonephrya complanata Ärnbäck-Christie-Linde, 1931

Body ovoid, spherical or conical, up to about 2 cm in diameter, but usually considerably less; surface with fibrils and adhering sand, etc., but not usually heavily coated; 7 small branchial folds on each side but the dorsal fold on either side may be reduced or absent; primary intestinal loop narrow, secondary loop widely open; one gonad on each side, the left one in the secondary intestinal loop; each gonad consisting of ovoid ovary and a group of male follicles around its posterior end; oviduct directed anteriorly. Larviparous.

On rock, stones, shell, etc., from the lower shore to depths of more than 500 m. Recorded from north, west, south and (rarely) east coasts of the British Isles but, because of its small size, may be more common than the records suggest. Further distribution includes the North Sea, European coasts, Arctic waters and north east North America.

Genus Eugyra Alder and Hancock, 1870

Eugyra arenosa (Alder and Hancock, 1848)

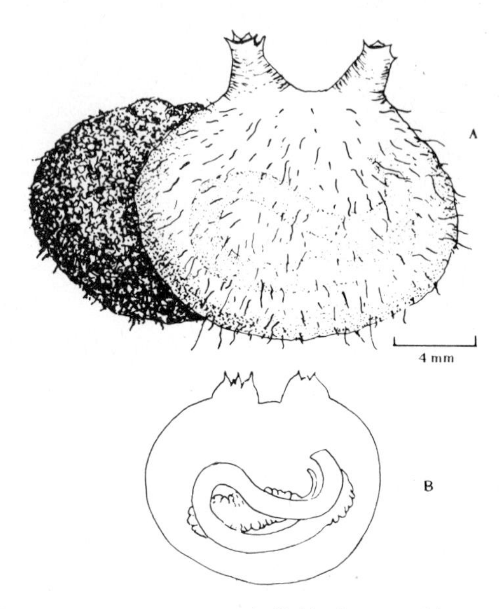

Fig. 60. *Eugyra arenosa*: **A,** two individuals, one with test cleaned; **B,** gut and left gonad.

Molgula arenosa Alder and Hancock, 1848

Body ovoid to globular, up to 2 cm in diameter; test with numerous fibrils to which mud, sand or shell adhere; siphons fairly close together; branchial sac without folds but with 7 internal longitudinal bars on each side; primary intestinal loop quite wide; secondary loop nearly semi-circular; one gonad on left side, none on right; gonad within and partly crossing secondary intestinal loop and consisting of tubular ovary and fringing male follicles. Oviparous but development is direct, without a larval stage.

Unattached, on sand or mud, from shallow water down to about 150 m. Generally distributed around the British Isles. Elsewhere, in European waters from the English Channel to north-west Norway.

References

This list includes references to some of the principal works on British ascidians but not to all of the original descriptions. The monograph by Berrill (1950) is the best account of British species and should also be consulted for much other information on the group as a whole.

BERRILL, N. J. 1950. *The Tunicata with an account of the British species*. Ray Soc., London.

CARLISLE, D. B. 1953. Notes on the British species of *Trididemnum* (Didemnidae, Ascidiacea), with a report of the occurrence of *T. niveum* (Giard) in the Plymouth area. *J. mar. biol. Ass. U.K.* **31**, 439–45.

CARLISLE, D. B. 1953. Presenza di spicole in *Diplosoma listerianum* (Milne-Edwards). Contributo alla sistematica degli Ascidiacea, Didemnidae. *Pubbl. Staz. zool. Napoli* **24**, 61–7.

CARLISLE, D. B. 1954. Notes on the Didemnidae (Ascidiacea). III. A comparison of *Didemnum maculosum*, *D. candidum*, *D. helgolandicum*, and *Trididemnum alleni*. *J. mar. biol. Ass. U.K.* **33**, 313–24.

CARLISLE, D. B. 1954. *Styela mammiculata* n. sp., a new species of ascidian from the Plymouth area. *J. mar. biol. Ass. U.K.* **33**, 329–34.

HARTMEYER, R. 1923–24. Ascidiacea. Zugleich eine Ubersicht über die arktische und boreale Ascidienfauna auf tiergeographischer Grundlage. *Dan. Ingolf-Exped.* **2** (6), 1–365; **2** (7), 1–275.

KOTT, P. 1951. *Corella halli* n. sp., a new ascidian from the English Channel. *J. mar. biol. Ass. U.K.* **30**, 33–6.

KOTT, P. 1952. Observations on compound ascidians of the Plymouth area, with descriptions of two new species. *J. mar. biol. Ass. U.K.* **31**, 65–83.

MILLAR, R. H. 1950. *Lissoclinum argyllense* n. sp., a new ascidian from Scotland. *J. mar. biol. Ass. U.K.* **24**, 389–92.

MILLAR, R. H. 1954. *Protostyela heterobranchia* n. gen., n. sp., a styelid ascidian from the Scottish west coast. *J. mar. biol. Ass. U.K.* **33**, 677–9.

THOMPSON, H. 1930–34. The Tunicata of the Scottish area, their classification, distribution and ecology. Parts 1–4. *Fish. Scot. Sci. Invest.* 1930 (3), 1–45; 1931 (1), 1–46; 1932 (2), 1–42; 1934 (1), 1–44.

Index of Species

Synonyms are shown in roman, the correct generic and specific names in italics.

Notes